Waste Away

Waste Away

WORKING AND LIVING WITH A
NORTH AMERICAN LANDFILL

Joshua O. Reno

UNIVERSITY OF CALIFORNIA PRESS

University of California Press, one of the most distinguished university presses in the United States, enriches lives around the world by advancing scholarship in the humanities, social sciences, and natural sciences. Its activities are supported by the UC Press Foundation and by philanthropic contributions from individuals and institutions. For more information, visit www.ucpress.edu.

University of California Press
Oakland, California

Library of Congress Cataloging-in-Publication Data

Reno, Joshua.
    Waste away : working and living with a North American landfill / Joshua O. Reno.
        p.   cm.
    Includes bibliographical references and index.
    ISBN 978-0-520-28893-5 (cloth : alk. paper)—ISBN 0-520-28893-9 (cloth : alk. paper)—ISBN 978-0-520-28894-2 (pbk. : alk. paper)—ISBN 0-520-28894-7 (pbk. : alk. paper)—ISBN 978-0-520-96377-1 (ebook)—ISBN 0-520-96377-6 (ebook)
    1. Sanitary landfills—Michigan—Detroit.   2. Sanitary landfills—Social aspects—Michigan.   3. Refuse and refuse disposal—Social aspects—North America.   I. Title.
    TD795.7.R46   2016
    628.4′4564097—dc23   2015029538

Manufactured in the United States of America

25   24   23   22   21   20   19   18   17   16
10   9   8   7   6   5   4   3   2   1

In keeping with a commitment to support environmentally responsible and sustainable printing practices, UC Press has printed this book on Natures Natural, a fiber that contains 30% post-consumer waste and meets the minimum requirements of ANSI/NISO Z39.48–1992 (R 1997) (*Permanence of Paper*).

CONTENTS

ILLUSTRATIONS

FIGURES

TABLE

ACKNOWLEDGMENTS

There are many people who helped make this book possible, including scholars and friends on both sides of the Atlantic, and research informants on both sides of a debate concerning where landfills belong and whose waste they should accept.

First and foremost, I owe a debt of gratitude to the activists, residents, and waste workers who welcomed me into their lives and shared their experiences of the landfill with me. Without the cooperation of the managerial staff at "Four Corners landfill," I would not have been given the opportunity to understand and observe life on the ground as a landfill laborer. And without the assistance of my former coworkers, who looked after me, confided in me, laughed at and with me, this opportunity would have amounted to nothing. I am also very grateful to residents of "Harrison" and "Brandes" for helping me understand local history and inviting me to join them in a political struggle. I have done my best to faithfully represent what we said and did together.

This research was funded by the Center for the Ethnography of Everyday Life (CEEL) at the University of Michigan, which was supported under the auspices of the Alfred P. Sloan Foundation. Through the leadership of Tom Fricke, CEEL played an important role in many graduate careers, including my own. It introduced me to other young anthropologists, including Britt Halvorson, Sallie Han, Cecilia Tomori, and Jessica Smith Rolston, who were willing to discuss early fragments of my fledgling project and helped foster a community of intellectual encouragement and friendship. Of CEEL participants, Pete Richardson is the one who first introduced me to the possibility of doing "shopfloor ethnography" and inspired me with his redoubtable passion and wit. Many other people at the University of Michigan provided critical feedback on my doctoral research and writing at important moments,

ix

including Arun Agrawal, Guntra Aistars, Laura Brown, Fernando Coronil, Erika Hoffmann-Dilloway, Karen Faulk, Severin Fowles, Larry Hirschfeld, Eduardo Kohn, Daniel Latea, Oana Mateescu, David Pedersen, Jessica Robbins-Ruszkowski, and Katherine Verdery.

The ideas in this book have benefited immensely from the insights of other waste scholars. Early on in my graduate career, Gay Hawkins and Sarah Hill both encouraged me to pursue research on waste when this was still a relatively undeveloped field, for which I am incredibly grateful. After graduate school, I was fortunate enough to work with the incomparable Catherine Alexander, who offered vital support as mentor, collaborator, and friend. Through the Waste of the World project, funded by the Economic and Social Research Council, she and Nicky Gregson provided funding that allowed me to distance myself from North American landfills and gave me the freedom to begin transforming my dissertation into this book. Thanks along these lines are also due to Francisco Calafete-Faria, Melania Calestani, Mike Crang, Silvia De Zordo, Romain Garcier, Zsuzsa Gille, Luna Glucksberg, David Graeber, Keith Hart, Casey High, Vincent Ialenti, Max Liboiron, Randy McGuire, Alan Metcalfe, Kathleen Millar, Robin Nagle, Lucy Norris, and Ruth Van Dyke for affirming conversations about my work and their own, as well as to Liviu Chelcea, Eeva Kesküla, Daniel Miller, Jessie Sklair, and Daniel Sosna for enabling me to present my ideas in new forums and develop them further.

Gillian Feeley-Harnik, Doug Holmes, Stuart Kirsch, and especially Webb Keane deserve special acknowledgment for their critical feedback on various versions of this book and their intellectual and professional guidance over the years. Earlier drafts of these chapters were also read by Siobhan Hart, Karen Hébert, and Sonja Luehrmann, and the final version benefited enormously from their generous comments. None influenced it more than Britt Halvorson, however, who suggested so many terrific ideas and edits that I eventually stopped crediting her in the notes for fear I might have to list her as a coauthor! Along with the invaluable criticism of Matt Wolf-Meyer and the impeccable editorial insights of Reed Malcolm, all the people I have mentioned made this a far better book than it might otherwise have been. Though I would love to claim otherwise, I am solely responsible for any mistakes that remain.

Finally, it is impossible for me to adequately express my gratitude to my wife Jeanne and our son Charlie. The dissertation was dedicated to them, and so is this book, because neither would exist if I did not have them in my life.

Introduction

THE UNITED STATES WAS THE WEALTHIEST and most powerful society of the twentieth century and remains so at the start of the twenty-first. It is also the most wasteful in world history. Every day, seemingly without end, a wide variety of waste is collected and hauled away from businesses, government facilities, households, and communities—some to be recycled or incinerated, but most sent to landfills located elsewhere. This story of wastefulness is by now a familiar one, but landfills remain misunderstood, even mysterious.

This book seeks to reconnect waste producers to our landfills—to show the many ways in which we are already connected without being aware of it. Some people (a relatively small number) are already well acquainted with landfills. The lives of those who do waste work and/or dwell in the places to which it is transported become entangled with waste collection, transport, and disposal in various ways. To understand their experiences, I conducted interviews with waste workers and activists and visited landfill sites in the United States and Canada from 2004 to 2007. At a large Michigan landfill I call "Four Corners,"[1] I also worked as a paper picker—a common laborer paid to keep things tidy and perform odd jobs. I not only wanted to know how waste had affected other people's lives—I also wanted it to affect mine.[2] For nine months I touched it, walked through it, breathed it in, and brought it home. Afterward, I associated with local activists opposed to Four Corners. I attended get-togethers, meetings, and actions targeting the landfill's controversial burial of Canadian waste. This international traffic brought long-distance waste transport and disposal to public attention; it therefore enabled some Michiganders to openly politicize the separation from waste that most of us take for granted as a presumed right. Documenting these different ways of relating to North American waste, I explore the unexpected importance that landfills have for us all.[3]

It might seem strange to study one U.S. landfill so closely. After all, any one landfill, even a very big one, only has room for a tiny slice of the *mass waste* of mass production and mass consumption. Type II landfills like Four Corners receive most of our municipal solid waste, the rubbish left on the curb for public collection and disposal.[4] From one perspective, such waste is merely the final act in a process of resource extraction and commodity manufacture that is related to, but ultimately much bigger than, mere household consumption. It is these other kinds of industrial activity, in fact, that are responsible for most of the waste produced worldwide.[5]

Examining all of this waste in its totality can be useful but may leave us none the wiser about what waste circulation does to us and for us. Within environmentalist discourse, in particular, there is a prevailing tendency to aestheticize and moralize our problems with waste. Portrayed as a dirty secret that must be exposed, North American waste is most often used to provoke concern about human exploitation and pollution of the Earth as a whole. This book explores waste disposal as a social relationship, and not simply as a form of environmental abuse. Follow the contents of your garbage bag, recycling bin, or toilet and they will lead you to people and places to whom you are unknowingly connected, but never to an abstract and impersonal Nature. This social relationship is part of a waste regime that implicates consumers and producers, industrial and municipal waste alike.[6] The challenge of this book is to demonstrate how landfills and waste producers are mutually associated, how something absent from our everyday lives can still shape our experiences and imaginations.

The first story I heard when I began my research was from the man who would later become my boss: Bob, the operations manager at Four Corners. During this conversation, we were in the white Ford F-150 he called his office, perched atop the bloated mountain of waste he had been constructing over the past twelve years, first as an operator, then as a high-ranking manager. Bob was married, approaching middle age, and had one stepson he loved dearly. Like a significant number of his employees, he grew up on farmland in the rural outskirts of the Detroit metropolitan area; but he had given up his dream of working that land, for a steady, middle-class income in the waste industry. Like all his employees at the time, he was white. I'd begun to ask Bob questions about what landfill labor was like and the challenges it presented that other kinds of work did not. In response, he decided to tell me about one of the worst days of his life.

Landfills are a lot like other construction sites, which means they can be dangerous places to work. After the large, transnational firm I call "America Waste" acquired Four Corners in 1999, the site developed a company-wide reputation for a low accident rate. But in the fall of 2003, Four Corners had its first accidental death. The man had been hauling waste for only two days, which is why many landfill personnel blamed the accident on inexperience. According to the police report, he had stood in the wrong position while preparing to dump the contents of his sludge tanker into the fifteen-foot-deep trench known as the "sludge hole." In the process, he was knocked head first into the trench and began suffocating underneath the thirty-five tons of processed Canadian sewage he had just finished hauling across the border. Another truck driver spotted the incident from a distance and signaled land-fill management. Bob described to me the events that followed:

> I come into work that morning and I was here about a little bit before six. I walked up to my office and they two-wayed me on my Nextel and said that, uh, somebody'd fell in the sludge hole. It was dark at that time so you know I asked 'em again because it kinda startled me and didn't sink in. I thought I heard what they said. So I ran downstairs, jumped in my truck, came fly-ing out back here. And of course we had light plants so I could see down in the hole and could see his legs stickin outta there. Uh, so I immediately, not thinking, I know at night the sludge hole isn't completely full so we put a layer of autofluff so that there's a little bit a layer to hold the stink down. I knew that was down there so I jumped down on that stuff so there was probably five, ten feet of sludge below me. I was standing on the autofluff and his legs were right there where I could reach them. So I grabbed hold of his legs but I couldn't pull at all so I grabbed a hold of his pant leg and rolled it up in my hand to get a hold of something to pull on, so I was pullin' on that, meanwhile a couple other guys came out there, they threw shovels down and couple other guys got down there with me and we tried pullin' on it. But there's like a suction in there once you're surrounded with it and there's no way we could get him out.

At this point, local police and firefighters had arrived. Some climbed in the sludge hole with Bob and the other workers and tried to help (one fireman would later check himself into a hospital to have his blood tested for infec-tion from the sludge). Bob got out of the trench and focused on removing the body: "I tried to collect myself a little bit. We dug him out with a backhoe and laid him out on a board down there. They told us we couldn't get him outta the hole, otherwise it'd be a crime scene, so we laid him on a board down there." Bob told me he stood there only a moment, covered with pun-gent, grayish sludge and looking at the dead body of a truck driver he didn't

FIGURE 1. The sludge hole.

know, before he realized the landfill was going into disarray. At the time, Toronto was sending all of its processed sewage to Four Corners, which made up the majority of the two thousand tons a day of sludge buried at the site, the rest coming from wastewater treatment plants in the greater Detroit area. Without a trench for the morning's incoming sludge, the flow of truck traffic was disrupted. "All the sludge trucks were parking out back, so we had twenty, thirty sludge trucks. We had to get the operator and get him on another backhoe so we could start digging another sludge hole."

Though order was soon restored, Bob said the experience gave him a glimpse of Four Corners that he'd never forget:

> It's one thing about this business: the garbage never stops for nothing. Trucks keep comin'. It made me think a little bit about it that time, if I died now this place wouldn't stop for nothing. Garbage is coming here no matter what, hell or high water. We've never closed down here for nothing and it's just the sheer volume. This place keeps still rollin'. You gotta figure out why it happened. It wasn't only his second load and he wasn't trained properly. But this place won't stop for nothing. The garbage keeps coming.

The first time I heard this story from Bob, sitting at the top of the landfill, he repeated this last line several times in succession, pausing in between:

The garbage keeps coming.
The garbage keeps coming.
The garbage keeps coming.

The repetition of this phrase partially served to emphasize the significance of what Bob was saying, but it also interrupted his narrative and shifted focus back to the context of our interview, as we sat there in his truck, watching some of those same waste haulers arriving from across the border, one after another, waiting for their turn to dump. I take "The garbage keeps coming" not as a moralizing statement about the wastefulness or the environmental destructiveness of North American consumers, but rather as a critical clue about the material conditions that make possible their particular way of life. This way of life happened to be one that Bob and I shared, but he did not say something to index our collective culpability either, such as "We keep sending garbage to the landfill." Nor did he try to hold Canadians responsible for the sludge that had claimed the man's life by saying, for example: "They keep sending garbage." Instead, Bob had made waste itself the subject of his sentence, suggesting through repetition that the movement of the garbage elsewhere is somehow inescapable. It was as if all this waste had a life of its own. For the same reason that it offered steady employment to Bob and his employees, the garbage could swallow up a truck driver and keep coming as if nothing had happened.

Why should Bob describe his job in this way? How does it happen that ordinary things we casually throw away take on new properties and meanings when collected together? How is it that we have become disconnected from our waste, such that we bear no responsibility for what becomes of it and whom it affects? Addressing these questions means getting closer to landfills to understand what they do and why. As many have noted, the movement of waste elsewhere creates a distancing effect—a separation between waste workers and waste makers—yet what goes on at landfills continues to shape our lives, behind our backs and beneath our notice. Waste is not only *made* by us; learning about the particular people who live with landfills (sometimes by choice, sometimes not) brings into relief the many ways that waste *makes* all of us, even as it is buried further and further out of sight.

The garbage that keeps coming, to which Bob referred, is specific to North Americans and to other people who treat waste the way North Americans do. In a sense, landfilling represents only the latest expression of a century-old revolution in engineering and governance that has progressively divided people from their waste. How we define waste is relative—different people and places evaluate materials differently—but equally variable is what people do with the things they consider waste, and that *difference that makes a difference* is the focus of this book. I characterize "the garbage" that Bob refers to and that I labored upon as *mass waste* in order to address the historical specificity of North American waste disposal and the particular way of conceiving disposability to which it leads. In this sense, mass waste is what we think about waste as a consequence of what we do with it, not the other way around.

When a man dies while trying to bury material that others have cast aside, why is it that none of them are held morally accountable? Part of the reason, perhaps, is that the sludges that come from wastewater treatment facilities bear little resemblance to what any one person sends down the toilet or drain. Modern waste-management systems physically transform materials as they move and gather elsewhere. This transformation separates sludge and other waste from our everyday lives and forecloses the possibility of further moral reflection. In other words, that no one is morally culpable for the death of a man drowned in sewage sludge is due to the very work that landfills and waste workers perform. By hauling the anonymous excrement of Canadians across the border, that doomed driver delivered the instrument of his demise and ensured its depoliticization as nothing more than an amoral workplace accident. In other words, it is not only that the long-distance transport of waste creates potential hazards where it comes to rest, but that this process limits our ability to imagine ourselves as connected to what we discard. How could we bear responsibility for the unforeseen consequences of waste disposal when few of us even know where our waste ends up?

Most of us are never confronted with substances like sewage sludge because of the labor of waste removal and disposal, which reduces the range of experiences and relationships we can develop with material things and with each other. Landfills have made possible the cheap and efficient separation of people from their discards, the absence of which changes our very ideas of disposability. The ideal landfill not only hides our waste from us, but is itself hidden elsewhere, designed to blend seamlessly into an out-of-the-way landscape. And increasingly, North American landfills are not only out of sight but out of state, as waste travels ever greater distances to privately owned mega-landfills like Four Corners. Waste sites remain relatively inconspicuous to all but those who know what to look for, or who live close enough that they cannot help but take notice. But mass waste disposal does not simply put things where people think they belong; it leaves behind a constitutive absence through this act of subtraction. The power of this absence over our lives becomes clear if we resist the infrastructural and ideological pull of waste materials out of our everyday orbit and immerse ourselves in landfills and the lives of people who work and dwell with them.

Such resistance and immersion are not pleasant. During my first weeks as a paper picker at Four Corners, I was always relieved to return home, where I could take off my filthy clothes, shower, and enjoy being removed from the potent sights and smells of landfilled waste. Eager as I was to begin my research, what I carried home with me, day after day, was disconcerting to say the least. Among other things, my early field notes are filled with descriptions of my inability to empty all of the dust from my wallet, my pockets, my bodily apertures:

> It gets everywhere. Used four Q-tips today and ears were still dirty. Dirt in my mouth from dust blown by trucks, by gusts [of wind], by small dust devils . . . it makes me spit quite often.

As I came to embrace my after-work purification rituals, I also began to worry that I was not getting close enough to the landfill. For one thing, my apartment was in the same county, but outside of town; I did not live on the landfill property, as two employees did, or across the street, as did one other current employee and two former employees. My home life remained unaffected when the open face of the landfill belched up odors or dust carried on the wind. My education in anthropology had taught me to seek out challenging forms of difference, and I worried that the temporary freedom from filth I enjoyed was keeping me out of touch with what it really meant to work at a landfill, which I believed meant being constantly confronted with waste.

Later I would learn differently. When I finally acquired a uniform and work boots, I began following my coworkers to the locker room after we clocked out. Then I realized that purification rituals were equally important for them as well, only they began at work. Soon I learned to change into fresh clothes and shoes at the end of my shift, leaving behind a soiled uniform to be laundered. As I discuss in chapter 1, many of my coworkers were just as invested in separating Four Corners from the rest of their lives as I had been, some even more so. There were those who lived even farther away than I, in neighboring counties, in subdivisions and small towns where they could conceal their occupation from their neighbors if they chose to (and some did). I discovered that not only was I similar to them, much to my relief, but that many of my coworkers aspired to benefit from the same separation from waste that their labor provided for others.

My sense of being distant from landfill activity was also illusory. My body and bathroom were routinely sullied and cleaned, over and over again. What maintained my bathroom was the application of chemical agents, whose industrial production had generated landfilled waste and whose final use would leave me with disposable containers. The soap and water that carried away the dirt from my skin, down the drain, was bound for a wastewater treatment plant whose processes of water purification regularly generate many tons of sludge, enough even to drown a man. Cleanliness is made possible because of all that is inconspicuously absent and landfilled elsewhere.

My ideal conception of cleanliness was not of my own invention, moreover, but historically specific to and constitutive of social life in modern North America. I would perform similar acts even had I not just come from a land-fill.[7] In English, to dump or dispose of people and things means devaluing and bringing to an end a connection you once had. What and who we reject, in this sense, helps us become the people we aspire to be. This was the influential insight of anthropological analyses of impurity and pollution during the 1960s and '70s. Mary Douglas, Louis Dumont, and Edmund Leach argued that the seemingly universal tendency for certain people and things to be classified as polluting or taboo was a byproduct of a human demand for meaning and order. The original thesis most associated with Douglas (1966) is that "dirt"—and, by extension, any form of filth or rejectamenta—is "matter out of place." After I finished a shift working at Four Corners, my goal was to remove dirt from my person, my clothes, and my home, where it did not belong. This dirt would not exist, as such, if it did not offend some all-too-human, moral sensibility about the way things ought to be. Understandings of cleanliness differ according to cultural and socioeconomic backgrounds, but we all, by virtue of being human, seek to impose some sense of order and predictability on our bodies and surroundings.[8]

This raises a seemingly simple question: Why is it that people and things always have to be recleaned, over and over again? If "order" refers only to culturally specific and uniquely human descriptions of—and prescriptions for—the world, why does the world seem so persistently resistant to order, regardless of what specific kind of order that is? This is the question a young

Mary-Catherine Bateson asks her father, Gregory, in a dialogue at the start of his book *Steps to an Ecology of Mind* (1972: 3):

DAUGHTER: Daddy, why do things get in a muddle?

FATHER: What do you mean? Things? Muddle?

D: Well, people spend a lot of time tidying things, but they never seem to spend time muddling them. Things just seem to get in a muddle by themselves. And the people have to tidy them up again.

The elder Bateson's answer is that there are far more states for things to be in that people call "muddled" than "tidy," which he uses to introduce the important cybernetic concept of *entropy.* Entropy is often defined as "disorder," but in classical thermodynamics it is considered a real condition of the universe, not an epiphenomenal and all-too-human gloss on material events. To avoid confusion, another way to define "disorder" is as a state of open-ended rather than limited possibilities (Deacon 2012: 228).[9]

The inverse of entropy, in this sense, is *constraint,* or "the property of being restricted ... less variable than possible" (Deacon 2012: 193). For the Batesons, tidying is a way of limiting the possible arrangements of things in the world, reducing them to states that are preferable. Let us imagine what would happen if the Batesons ceased tossing and flushing and no one came to remove their waste. After a time, their home would gradually transform. Decaying food and bodily waste would break down and attract various nonhuman creatures, what we normally think of as pests, seeking nourishment and shelter. Increasing clutter would afford these invaders more opportunities to conceal themselves and build nests. Moreover, if elemental forces were allowed to act on the house and they never made repairs, it would gradually lose structural integrity as well, walls and floors would deteriorate, pipes would rust, furniture would mold. From the perspective of the Batesons, the house would indeed appear muddled and barely resemble the structure it once was, but it would have provided new growth out of decay. By tidying, cleaning, repairing, and refurbishing buildings, we actively constrain the kinds of places they might otherwise become. As a result, our offices, hospitals, bedrooms, bathrooms, classrooms, and factories continue to last as the material settings they are from day to day.

But what does it mean for something to last? After all, there is a sense in which the same thing cannot exist over time because everything is always in the process of changing into something else.[10] As Bateson argued, entropic change is irresistible for the simple reason that there are just too many states for things to be in that are disorderly. And yet, the keyboard in front of me, my hands typing on them, and the room that now surrounds us . . . each of these things perdures from moment to moment, more or less independently from each other, from my conscious reflection, and from the specific molecules they happen to consist of. Any arrangement of matter that lasts in this way can be called a "form."

Form, in this sense, can be contrasted with processes of (de)formation or becoming.[11] When any actual thing does persist over time, whether a commodity, my body, or an enclosure, it is because its specific form withstands change in some way. And whether or not we are aware of it, many of our routine actions halt or slow down process and change. The durable form that now surrounds me—and, in a sense, is me—is sustained through active subtraction: dusting and replacing computer keys, shedding and regenerating skin cells, and regularly renovating and tidying my home. Most of the time, what we mean by "waste" is a necessarily unnecessary product of creating and maintaining form—an expenditure that is lost so that durability can be gained, even if only for a short while.[12]

One simple way that we stave off the deformation of our things, places, and bodies is by encasing them within relatively durable structures, such as buildings. These structures are never finished, moreover, but are regularly reconstructed or altogether replaced to remain stable containers for sameness. Sometimes, Four Corners would receive recently abandoned mobile homes still stuffed with the possessions of prior occupants. On one occasion, I stood atop the landfill and watched with Eddy, a brash younger laborer, as a double-wide trailer was dragged to the dumping area, carved up by a bulldozer, and pushed into the sludge pit. We witnessed the accumulated memories of a household spill out of the structure—furniture and photos and clothing—as all were shoved into the liquid muck below. It seemed the former occupants had chosen to leave these items behind rather than make further payments on the lot, or they were forced to. The rentiers were likely concerned about maintaining the value of their property, which was the land such homes rested upon and not the enclosures themselves.[13] If the trailer were allowed to remain, it would likely attract unwanted human and nonhuman squatters, making the property harder to resell. By quickly and efficiently removing and burying unwanted trailers, mass waste disposal helps that land to last in a state of relative stability that forecloses other possible uses and limits its availability to paying customers only.

By underwriting our endless tidying and cleaning, mass waste disposal allows us to actively avoid alternative possibilities we would rather not experience or imagine, such as disused buildings overrun by pests. Landfills accomplish this in a way unlike any other disposal method, however, and this exerts a powerful influence on our lives. The availability of simple and cheap separation from our waste has, in turn, made possible new kinds of disposability, allowing yet more eventualities to be carefully avoided, in the service of private and public interest.

Here appears our first apparent paradox. How could an attachment to sameness, to things that last, coincide with widespread disposability, which implies a lack of attachment to things? When it comes to our bodies and dwellings, it is easy to see how routine material expenditures provide them with enduring form. But what about the commodities we routinely consume and dispose of? Every specific iPad charger, bottle of Coke, or viewing of *American Sniper* may last only a short while, but each ideally provides the consumer an identical physical experience with a mechanically reproducible form. Yes, the charger will malfunction, the soda pop will eventually go flat, the digital movie file will become corrupted as a result of entropic deformation ... but you can just go buy an identical replacement. The perpetual reconstruction of sameness makes possible a consumer's attachment to a suprasensible value that transcends each momentary encounter with an individual purchase.[14]

And, as I discuss in greater detail in chapter 3, for commodities to satisfy our demand for reproducible sameness, abundant waste must be created. For example, Four Corners would occasionally receive the smoke-damaged contents of a liquor store (known as a "party store" in the Midwest). Agents from the Bureau of Alcohol, Tobacco, Firearms and Explosives (ATF) would escort the contraband to the dumping site and watch, guns holstered, as operators covered the loads with garbage, enough to satisfy our armed visitors that the spoiled spirits were no longer available for public sale and consumption, but not enough to prevent operators from digging it out once they were left alone. The government's involvement can be understood as a way of ensuring disposal in order to avoid unwanted possibilities, specifically the reuse of potentially harmful liquor and a threat to corporate liability. If waste collection and disposal constrain the range of experiences we are likely to encounter, this includes limiting our opportunities to scrounge for goods to use or recirculate. The absence of smoke-damaged liquor protects consumers from unnecessary risk, but it also promotes the purchase of alcohol new and, ideally, ensures that each drink will be the same as any past or future drink of the replicated commodity form.

As with refurbishing or demolishing houses or spoiling unsold alcohol, North American waste management choreographs a redistribution of possibilities such that things can be made to last in one domain by banishing the unruly elements subtracted from them. Sending waste away is a means of transcending inevitable process and change, however temporary or long-lasting this effect may be. But continually reconstructing sameness comes at a cost. From a cybernetic perspective, an organism lasts only by "continually sucking orderliness from its environment" (Schrödinger 1944: 73). This is another way of saying that there is only so much form to go around: generally speaking, the more things we produce, the more we cling to both newness and sameness, the more waste we produce and the more unstable the world around us becomes. The chapters that follow demonstrate that wherever mass waste goes, it destabilizes environments, values, social relations, bodies, and lives in open-ended ways. The reconstructed sameness of our everyday life depends on sucking stability from the lives of people elsewhere, and in contemporary North America this has been accelerated by the perpetual rises and crises of industrial capitalism.[15]

The elimination of waste exists wherever there is life, which depends on a separation between a lasting bodily form and routine expenditures of transient solids, liquids, and gases. But North American landfills pursue this to an extreme degree, which is why they have contributed to the creation of the most wasteful society in history. To understand how this happened, one has to explain how making things last became embraced as a way of transcending the world entirely, rather than merely of surviving within it.

Four Corners is known as a *modern sanitary landfill*. "Modernity" and "modern" are notoriously difficult to define, precisely because they are not absolute but relational terms. Typically these words are deployed to separate some groups, forms, and practices from those that are considered classic, primitive, traditional, and/or passé by contrast. A modern X is thus thought to have come after, and somehow transcended, its nonmodern counterpart.[16] For hundreds of years, people who considered themselves modern have demonstrated such transcendence in part by showing themselves to be free of filth, decay, and disease—that is, free of signs of entropic process and change. From the Renaissance onward, European elites increasingly positioned themselves against the carnivalesque rituals and uncivilized behaviors of the lower classes. To conceal one's grotesque animal nature, for example, disposable objects like napkins, tissues, or toilet paper can mediate the removal of unsightly bodily expenditure. The result was social and political transformation born of the mutual differentiation of unequal groups.[17] European colonists brought these productive analogies between human kinds and material forms with them to the New World. Furthermore, as I discuss in chapter 5, North American waste disposal continues to serve as both practical enterprise and playful idiom for social differentiation.

Today, having reliable waste removal and disposal services has become a measure of modernity the world over. Among the things that people in so-called developing (and hence nonmodern) countries typically lack is sufficient infrastructure to provide their people with adequate sanitation and access to clean water.[18] Sanitary engineers in North America routinely describe contemporary landfills as "modern" to compare them favorably with the common dumps of poorer places and a previous era. The first closed tips and landfills sought to standardize dumping in order to constrain what could be done with waste by humans, pets, and pests. Waste pickers, pigs, dogs, flies, fires, disease, explosions, and odors were all very real possibilities associated with the lifeways that waste disposal fostered in Euro-American cities prior to the rise of sanitary engineering in the nineteenth and twentieth centuries.

The first incinerators and landfills represented a shift away from waste disposal as an open-ended, multispecies enterprise and toward rational and technically sophisticated management. Prior to their introduction, North Americans had experimented with many ways of reusing waste. By the 1880s, many communities used a process called "waste reduction," also known as the "Merz process," which treated waste in order to produce fertilizer and grease. Many rural communities used hog-feeding methods well into the Great Depression. This declined when widespread cases of trichinosis in the United States were traced back to contaminated pork. Reduction also fell out of favor in the early twentieth century because it could not dispose of ash and other kinds of nonbiodegradable refuse, for which more expensive incinerators were needed. Landfills were cheaper and did not discriminate between types of waste—it did not matter whether materials could be recovered as scrap, grease, fertilizer, or hog feed. As the sanitary landfill rose to prominence during the postwar era, alternatives fell out of use. Arguably, this has only made the biosocial relations involved even stranger and less predictable.[19]

Through modern sanitation, wasted things are transformed twice over: thrown into a mass waste stream, each particular discard begins to lose its symbolic and material identity and, thus, its connection to the particular people, places, and things to which it was formerly attached. Once detached, it adopts a new identity as part of the waste stream. Whether it is recycled, burned, or buried, mass waste partakes of a collective future wholly separate from its multiple past lives. Because of its alienation from the people, things, and places that produce it, mass waste acquires an indeterminate character— anything could be in the waste. But there are different ways of imagining such possibility. On one hand, this indeterminacy is attractive to those who would scavenge mass waste for usable and resalable items: anything could be in the waste! On the other hand, such uncertainty represents a threat that must be tamed by the stochastic techniques of risk assessment and mitigation that characterize modern environmental governance: anything could be in the waste! The modernization of waste management in North America has typically meant that the latter form of indeterminacy takes precedence over the former—mass waste is feared for the threat it poses to the health and well-being of communities and environments. Landfills may be durable, but their contents are highly unstable, as are the resulting relationships they develop with their surroundings. The movement and deformation of mass waste multiplies opportunities for profit and protest in its wake—animated by different ways of interpreting the idea that anything could be in the waste.

Given its relatively indeterminate status, mass waste is characterized by the fact that it must move and gather elsewhere—a material imperative of state-craft that routinely subjects a few people to the unstable waste of many.[20] Some bodies, things, and enclosures are routinely exposed to the open-ended possibilities of mass waste on our behalf. To see waste as a social relationship means recognizing the subtraction of unwanted material from our lives as a form of care provided by others. They care for us by absorbing the unstable risks and benefits that mass waste proliferates. As David Graeber writes, "It is sometimes said that the central notion of modernism is that human beings are projects of self-creation. [W]e are indeed processes of creation, but . . . most of the creation is normally carried out by others" (2007: 100). Sanitary landfills are also modern in Graeber's sense—they intensify the reliance of North Americans on others who make possible their continuous self-creation, while hiding this reliance from sight.

In the first three chapters of this book, I focus on landfill labor as an example of modern care, considering both the dilemmas it creates for those who do it and the way it productively constrains possibilities for us all. Routine exposure to waste can result in stigma and suffering: throughout the world, "waste" is just as often used to refer to unwanted people as to things. Waste workers may worry about being mistaken for the object of their labor, seen as mere human waste. Yet, to be considered modern, North American waste work must involve machinery, mechanical skill, and higher wages. As a result, it has tended to become paradoxically stigmatized (as filthy) as well as privileged (as white and male), the effects of which are evident in the lives of contemporary waste workers.[21]

Such tensions are particularly acute for men in southeastern Michigan, where my research was primarily based. As the consolidated power of global capital has eroded opportunities for farm, factory, and construction labor in the region, waste work became an attractive substitute for some. This creates difficulties for which middle-class transcendence is seen as the cure. My coworkers' projects of self-creation typically involved having a home, a body, and possessions that not only last but will appreciate over time. This is precisely what landfilling makes possible, as a type of care from which many benefit— and what it can also threaten, as a source of wage labor that a very few struggle to earn.

Besides landfill workers and truck drivers hauling in waste from great distances, Four Corners received regular visitors from the surrounding communities, hauling bulk waste to dump at the "Citizen's Ramp": two roll-off boxes with a cement ramp in between them, located near the entrance to the site. Residents and local businesses could drive in, park on the ramp, and toss any items they could not dispose of through regular trash pickup. At regular intervals, the roll-off boxes were carried to the top of the landfill and dumped with the rest of the incoming garbage. Laborers were routinely sent to clean the Citizen's Ramp, and this presented opportunities for scavenging as well as observation. Among the most common materials to be found at the Citizen's Ramp were leftovers from the construction and renovation of homes, including old shingles, scraps of wood, and dilapidated furniture and appliances.

My fieldwork took place right at the peak of the U.S. housing bubble that would eventually precipitate the global financial crisis of 2008. At the time, construction was a major source of employment in Michigan and the leading source of income for residents of Harrison, the town where Four Corners was located; it was also the former occupation of many landfill workers. What sustained this industry—and, by extension, what constituted one of the central premises of global financial transactions—was the postulate that houses would continually grow in value at a predictable rate. Just as the early

Calvinist settlers described by Max Weber paradoxically worked hard to earn the salvation that predestination had already ordained, contemporary North American homeowners work hard to build equity through constant repair and refurbishment, despite a widespread belief in the guarantee of appreciation. Such activities account for one of the largest waste streams of modernity: buildings remade, newly built, or left to ruin. During the housing bubble, and still today, stripping away old materials and adding new ones is meant to secure the value of homes in perpetuity.

The middle-class house is perhaps the greatest expression of reconstructed sameness, of a form made transcendent through disposability, in contemporary North American capitalism. The logical endpoint of this middle-class dream was already apparent not far from Four Corners, in the depopulated and bankrupt city of Detroit. Today, nearly a quarter of the city's housing stock has been abandoned, totaling over 70,000 units. This is not only a potent symbol of the "white flight" that followed the riots of the sixties; it is blamed for lost economic investment, routine squatting, fires, vandalism, and violent crime (Kahn 2011). And yet here, too, a similar process of renovation through wasting is occurring on a grander scale. Demolition of homes has been, since before the housing bubble burst, actively politicized as a way of restoring or saving Detroit from ruin. By renewing their material foundations and eliminating unwanted possibilities through subtraction, houses and cities are made to last. Maintaining form means having a future. As I discuss in chapter 4, moreover, the telling and recording of histories is reliant on form. For places to have historicity, for "what has happened" to be recordable and narratable, traces of the past must be available to link up with meaningful futures. The township I call "Harrison" faces an uncertain future, not only because it has an active landfill, its second, but because it was the kind of place to attract one to begin with, a place ripe for preemption by the designs and desires of other people and places.

For anyone or anything to be made and have lasting form requires resources and effort. If that seems counterintuitive, it is partly because many North American moderns now take for granted the labor of waste disposal. Studying landfilling shows us that the redistribution of waste elsewhere helps make transcendence and transience salient features of everyday life in North America. By calling landfilling a social relationship, my aim is to remind the reader of the active production of transcendence out of sight, as well as the material imbalances that inevitably result. It is a small minority who, like Bob, manage transience on behalf of others. Understanding their lives brings into stark relief the absences that landfilling so effortlessly provides for the rest of us. This is what makes exploring life (and death) at Four Corners landfill in the chapters that follow worthwhile.

We all encounter waste every day—excreting, cleaning, taking out the garbage—and we all anticipate and enact separation from our waste in some way. Waste relations at and around Four Corners vary from these ordinary experiences not in kind, but in intensity and duration. As my daily rites of purification suggest, waste workers experience both immersion in waste and transcendence from it more intensely. Rather than employ *cross-cultural comparisons* (the traditional hallmark of anthropology) and thereby invoke notions of irreducible incommensurability, this book documents *transubstantial homologies* between life lived in and out of a landfill's orbit. That is to say, I focus on similarities rooted in relative immersion in, and separation from, transient substance.[22] At Four Corners, once our work boots and uniforms were beyond reuse or we acquired replacements, they were usually thrown away on site, thereby collapsing the complicated circuit by which most discards reach their destination. Similarly, when laborers finished bagging litter on site for the day, to be picked up during the night shift and hauled to the top of the open face, we would sometimes throw away our gloves in the last bag we filled. This act not only ritualistically ended our shifts, but allowed us to conspicuously partake of the convenient disposability that landfills inconspicuously provide for all. As I discuss in chapter 3, however, we would occasionally disrupt these circuits of disposal and scavenge things from the waste as well.

My coworkers' efforts to separate themselves from waste represent an amplified version of ordinary experience. What North Americans tend to share in common, whether we work at a landfill or not, is that landfilling actively makes possible our separation from some things and experiences while sustaining our connection to others. The movement of waste workers from Four Corners to their homes and back exposes them to collective waste so that most of us never have to know landfills. In this book, therefore, landfill workers are not presented only as people with a difficult and unusual job. Like us but unlike us, landfill workers are different in ways that reveal suggestive homologies. To the extent that the observations and experiences of waste workers accentuate and distort the separation from waste that most people take for granted, they can help illustrate how landfills orchestrate our material lives even (and especially) as they are more and more absent from them. If many North Americans have thinner or less intense relations with waste by comparison, examining the lives of people at Four Corners helps us appreciate these relations in a new way, as something that people like Bob make possible.

Our common reliance on waste labor and infrastructure need not lead to despair, moreover, but can be a starting point for us to reimagine environmental action on the basis of our shared dependence on *both* our natural environment *and* the care of others. Understanding the contribution that waste workers make to our lives, that their activities are more than a symptom of our reckless consumption of the Earth's resources, it becomes clear why a form of waste disposal so disliked is still widely used, why it is we cannot seem to do without landfills.

Interlude

A NOTE ON DRAWING AND ETHNOGRAPHY

IN PLACES LIKE FOUR CORNERS, the fact that the garbage keeps coming ensures regular exposure to anonymous streams of mass waste. Experiences that affect one on such a bodily level challenge the conventions of literary representation. According to Roland Barthes (1989: 321), writing has no smell. While this could be taken as a reflection on the limits of representation in general, for Barthes speech *does* smell. Speech outlasts writing in a way, hanging in the air like bad breath that clings to speaker and hearer alike. I take Barthes to mean that different representational modalities have different limitations and capacities, which could also be taken as an invitation to experiment with mixed representations, as others have done with mixed methods.

Initially, I had hoped to supplement the text of this book only with photographs depicting the events and people it describes, but the conditions of my fieldwork made this difficult. I did not feel comfortable bringing a camera to work with me regularly, where it could easily be damaged or get in the way of my labor, and often I would have only a cheap camera phone when the opportunity to capture an important moment presented itself. Then there are the limitations of photography (no doubt related to my limitations as a photographer). It is sometimes said that drawing, as opposed to photography, can make subjects seem more lifelike. Because drawing amplifies some actors and actions and excludes others, it subtracts from the total visual array available to a light-capturing machine. Arguably, this very selectivity more closely resembles the distorted parallax of the visual field our bodies perceive. While a human agent must frame photographic shots, moreover, the resulting image is mechanically mediated—the very movement of a writing implement across a page betrays the organic nature of the hand, which is evident in the traces it leaves behind.

As Erik Mueggler (2005) reminds us, all forms of representation stage and screen the surrounding world and its inhabitants in some way. Though land-fills shape and are shaped by people, some representations can make them appear radically separate. Whatever their other benefits, photographic images of mass waste can inadvertently focus the eye on abundant materials and imposing machines, making it harder to see the people involved in these scenes or to recall the mediating influence of the photographer and their equipment. Arguably, forms composed of the same ingredients of graphite and ink, yet drawn to appear distinct on a page, express both the artifice of discontinuity and the underlying reality of continuity in a unique way. The drawings I have included are meant to suggestively restage figure and ground, to show people in the orbit of a landfill as active agents, rather than portray them as the passive recipients of modernity's ruins.

Tim Ingold argues that drawing, far from an exercise in artifice, can offer traces that resemble the world as lived:

> Through the close coupling of perception and action, the draughtsman—like the walker—is drawn *into* the world, along paths of observation, even as he draws it *out* in the gestures of description and the traces they yield. Thus the drawn line moves *forward* in tandem with the movement of our own attention. To draw, as the artist Paul Klee famously observed, is to take a line for a walk. . . . The line is not an object or an image that we leave behind. It comes with us. (2011: 17)

The very word *drawing* is both a verb and a noun. If the depiction a drawing offers is more lifelike, it is not because of designs imposed by the human mind upon paper; rather, drawing unifies observation and description, the world and its representation.

Left with unusable photographs of important moments and people, I decided to redraw them, denying their disposability and making them last in a different way.[1] The results are four drawings of different kinds, arranged throughout the book so as to complement the writing as an additional modality of ethnographic representation. They are not meant to be aesthetic breaks from the textual narrative, nor mere prosthetic extensions of my argu-ments. More than a practical solution to the material conditions of my par-ticular fieldwork, drawing is an opportunity to reconsider ethnographic representation more broadly. The work of hand–eye coordination, of moving through and being with a place that drawing brings together, uses the very same capacities that enable many other forms of "participation" in a setting. Leaving traces is part of conventional fieldwork and of note taking. Many of my field notes were written by hand in notebooks; as I walked the landfill in the midst of daily tasks, I quite literally took "lines for a walk" as I worked and researched. More than the imposition of a "perspective" or "subjectivity" on the world I explored, my notes were directly shaped by the landfill, which left additional traces of its own in the form of questionable dust that fills the air to this day whenever my notebooks are opened. The notes therein are not merely ideational and representational, but also haptic and *fleshed out:* I still have calluses from work I performed at the landfill years ago, just as I still have a pronounced callus on my middle finger, just below the nail, where then, as now, I rest implements to write or draw. In this sense, drawing also speaks to a fusion of body and place (discussed in more detail in chapter 1).

Thus, my use of drawing is not an additional theory or perspective to include in the ethnography, but another way of *situating* my account. My hope is that the mediation of specific semiotic modalities (writing, observing, drawing) do not take the form of an arbitrary pastiche, but resonate with each other as differently situated forms of participation in a setting. At the same time, I selected these drawings to bring a hint of serendipity to this ethnography, as was present in the research process. This is something not often discussed in ethnographic writing—that there is something about going and doing that allows for unexpected things to happen. Like a drawing whose eventual form betrays its own process and history of formation, ethnography always bears signs of the possible and serendipitous. Getting into the mix is a hallmark of it as a method.

Drawing from photographs loses some of the serendipity of drawing from life, but imitation is no less a form of invention. The act of drawing takes on a life of its own through kinesthetic rhythms of the hand's own making, but also that of the page, the implement, the materials, the available light. Ethnographic drawing brings to bear the real possibilities and constraints of a different semiotic modality from writing, one that does not more authentically reflect "being there" or "knowing that place" or the people. If anything, it just gives my hands and eyes an opportunity to share more of what they know. This is only fair, important as they were for getting me through my time as a paper picker at Four Corners.

ONE

Leaky Bodies

YOU DO NOT NEED TO KNOW MUCH about landfills to know that you are not supposed to like them. But why should this be? After all, they provide an important, if regrettable, function. North American landfills are characterized as "sanitary" because of the historical legacy of public health management that sought to divide general populations from the filth they produce. And yet, the most common criticism of landfills is that they leak. Whether they *must* leak or simply *may* leak is a matter of dispute, but the possibility that they could fail to deliver on their promise of containment is what most North Americans know about them, if they know anything at all.

Maude was one of the first people I met during my fieldwork who expressed a strong dislike for Four Corners. The first time we spoke, she suggested that if I wanted to understand how the landfill was affecting the people of Brandes, the village where she worked and lived, the wider township of Calvin, and the surrounding area, then I should test the ditches along the county road for contamination. "Either that," she said, "or you should do a survey of the people who live around here and ask about their medical problems, because *a lot* of babies are being born with acid reflux." She later repeated this and other allegations:

> The acid reflux . . . every baby I can think has been born with it. Everyone is on medication for it. Anne's grand- son, Anne's got it, and Bill's got it.

> I never did have [allergies] but, you know, now I do have problems with my sinus and I never did have . . . all of a sudden I'm thinking, "My God, this is terrible!"

> You know, is it causing cancers? And I know people die from cancers but it's been a lot.

A middle-aged, white Michigander and grandmother, Maude worked along-side her sister-in-law Gwen at Lions Service, a party store and towing opera-tion in Brandes. The county road was the main thoroughfare connecting the interstate and the landfill, and dirt and debris often accumulated along it from passing waste haulers. It was along this road that Maude grew up. The site originally proposed for Four Corners was in the woods across from Maude's childhood home, and her politically active family had helped oppose this plan. Their victory was short lived, for a replacement site was chosen just down the road, in neighboring Harrison. Fourteen years later, working at Lions Service every day, Maude wondered if she were bearing witness to the landfill's impacts on local health.

Maude's claims highlight a paradox concerning the connection between landfills and bodies. Modern waste-disposal technologies were actually designed to *improve* the lives of people in crowded settlements; more specifi-cally, they were invented to help shield bodies from contamination. When large numbers of people live near one another (along with domesticated ani-mals and pests), parasites spread easily, resulting in frequent illnesses and occasional epidemics. Burying mass waste out of sight is not the sole means of avoiding these unwanted possibilities, but it does make possible homes and neighborhoods that are relatively free of pestilence, no matter how crowded they are. Insofar as mass waste removal helps people last, it represents a quin-tessential form of biopolitics, or the power to make individuals and popula-tions live or let them die (Foucault 1997).[1]

This power over vitality is easy to take for granted when mass-waste infra-structure operates smoothly. With the rise of bacteriology and the germ theory of disease, the profession of sanitary engineering was increasingly separated from the administration of public health in Euro-America.[2] As a consequence, the removal of waste has seemed less and less about bodily health, and more and more about environmental hazard. The rise of the envi-ronmental movement in the late twentieth century renewed anxieties about the unseen distribution of risks associated with modernity. Since at least the Love Canal disaster, it has been widely recognized in North America that landfilling does not eliminate environmental and health risks entirely, but concentrates them elsewhere. The possibility of leakage dominates how waste sites are currently regulated and run. Landfilling is thus both product and object of biopolitical control. Maude is not an epidemiologist or an environ-mental historian, but she knows to be wary of the invisible threat that Four Corners could pose to her home and workplace. She has good reason to read the bodies around her for inauspicious signs of illness, to look for patterns. This is common among people who live next to landfills, though the risks of waste sites have the potential to reach much farther.

In this chapter, I discuss how people work and live with the unruly possibilities of places like Four Corners. These possibilities are manifested in the form of real and imagined leaks—the landfill's actual and virtual seepage into the surrounding environment and into their bodies. For such leaks to affect living things, human or otherwise, presumes a breach in containment at two ends. The landfill must have an opening of some kind that allows materials to escape, and living bodies must have apertures that can absorb them. Whether one's home and health are threatened by such leaks or one is paid to control and monitor them, they are registered as bodily affects.

Discussing how North Americans are challenged by proximity to landfills calls into question the ways in which their bodies would otherwise be related to their surroundings—not only what disrupts North American ideals of body–environment relations, but what makes them possible. An animal's body is in perpetual contact with the beings and settings that surround it; its leakiness is paradoxically necessary in order to maintain form.[3] According to the "hygiene hypothesis," the pursuit of sterile hygienic surroundings may increase the risk of certain ailments, acid reflux among them—in other words, separation from waste, no less than exposure to it, can affect the growth and development of bodies.[4]

However, North Americans tend to believe that one body is intimately and exclusively associated with one person and that both are formally distinct from other beings and from their mutual surroundings. A body that is whole and singular, rather than divided, interconnected, or leaky, is an ideal they aspire to. For similar reasons, the loss of bodily fragments like fingernails and hair was once a matter of great concern among European Christians of the Middle Ages. How could a body in pieces, a body left to rot, or a body cremated be restored through the foretold Resurrection? Thomas Aquinas is credited with offering an Aristotelian solution to this medieval dilemma. The soul inhabits the body during life, but the latter is only accidental and transient substance, the temporary home for an eternal form. Bodies may provide a temporary location for personhood, but one's true essence is something more than the material world (see Bynum 1991, Rabinow 1996).

Medieval Thomism was later enshrined in modern secular institutions. In medical discourse and practice, for example, a whole and fixed patient/body is cared-for through the simultaneous sterilization of medical instruments and settings and the affective labor of bedside manner among practitioners. However, even here the body's wholeness and sameness are only virtual at best.[5] The continual elimination of biomedical waste and the purification of these sites is an essential component of the patient/body's lasting form across multiple examinations and treatments. Only the fast-food industry produces a greater volume of waste in North America than the medical establishment, which eliminates its waste not only for the enduring health of the patient but to protect medical institutions from legal liability.[6]

By the eighteenth century, regulatory regimes were gradually being imposed throughout Europe and its colonies to cope with the Industrial Revolution and to prevent the epidemics and fires of previous centuries. As it does for medical practice, North American mass waste management allows the occupants of many homes, businesses, and stores to hang together, or last, by constraining their surroundings. In indoor spaces, a state of air-conditioned neutrality prevails, ideally keeping out odors, the elements, and creatures that do not belong. The resulting absences further conspire to promote a sense of private autonomy and subjective freedom in North American dwellings—most obviously, perhaps, through the compartmentalization and concealment of excretion and other cleaning practices. Conceptions of acceptable cleanliness and orderliness changed with the implementation of new waste-management practices. The techniques and materials of waste management are not just a response to previously imagined ideals—they also constrain the imagined body/person in turn.

Having isolated and expelled all signs of transience, a regime of intensive sanitation radically constrains our immersion in everyday settings. In this way, waste-management infrastructure unmoors and individuates free selves from their surroundings.[7] On one hand, space can now be perceived as if it were divisible into discrete units that can be owned and occupied in a detached and disembodied way. On the other hand, bodies can become objects of individualized discipline. In liberal democracies, specifically, health and wholeness are increasingly imagined as the responsibility of self-creating subjects. The general promotion of hygiene, exercise, and healthy diet exemplifies this "government of the self, by the self." Whether or not they live in proximity to landfills, North American ill tend to be troubled by the gradual loss and deterioration of bodily tance, just as Maude is. The valorization of bodily sameness and perfect goes far beyond the public health establishment and includes cosmetic ies as well as the general social prejudice in favor of the young, healthy, fit.[8]

Most North Americans worry far more about aging and illness than about landfills, though the latter can amplify concerns about the former. Leaking landfills challenge the aspirational separation of bodies from surroundings and the freedom for self-creation this entails. In a landfill's orbit, bodies mix with and move through places and materials in challenging ways. This mixture of bodies and landfills is under close regulatory scrutiny, though not usually in a way that would satisfy Maude's epidemiological concerns. To explain landfill regulation, I begin this chapter by depicting the contemporary sanitary landfill as a disciplined and self-governed body, one that is cordoned off from (and displayed to) others with the assistance of technical expertise and labor. Landfilling transforms whole horizons, and landfill companies and regulators seek to control the leakages that result. As I will show, however, attempts to make Four Corners satisfy regulatory standards are routinely frustrated. This has as much to do with the uncertainty of landfilling as it does with the corporeality of being a landfiller. Evidence of a contained landfill body is partially embodied by managers and regulators, whose situated perceptions are meant to realize the standards of environmental governance. The human body is no less complicated to manage than that of a landfill. This comes across in the difficulties of life on the ground as a common landfill laborer—those workers who are most physically immersed in the landfill landscape. I discuss their distinct travails on the basis of my nine-month stint picking garbage on foot.

By describing landfills as bodies, I hope to complicate the presumed artificiality and haphazardness of the former while also suggesting that the mixtures they proliferate are ultimately beyond human control. Landfills are body-like because they involve the management of leaks and the circulation of fluids, basic requirements for all animal bodies irrespective of size, sex, or species.[9] The North American waste industry is organized around the possible threat of contamination; landfills are regulated to limit exposure to leaks; and, ideally, this allows mass waste disposal to fade into the background, unnoticed. My corporeal metaphor helps explain what makes landfills different from routine dumps, and the different types of workers involved in growing them and managing their leaks. But in some ways this is more metonym than metaphor. We are familiar with thinking of bodies as self-organizing systems, with innards and organs, but there are many such systems beyond the realm of living beings. Landfills can grow only by combining with other dynamic, self-ordering systems in novel and emergent ways. Microorganisms also dwell inside the guts of landfills, cousins to those helpful species that reside in our own guts. And because of this invisible feast occurring in their bellies, landfills also fart malodorous, combustive biogases: a source of potential energy, and inspiration for local odor stories and formal complaints to regulatory agencies.

More generally, remaking the surface of the Earth interferes with the ongoing production of atmospheres, aquifers, and soils. This occurs at that point of contact where the lively surface and the gyring sky blend with, and continually recreate, one another. It is through their dynamic interaction, through the inconspicuous productivity of dirt, that oxygenic atmospheres may be regenerated or suffocated with carbon emissions, and aquifers recharged or polluted. A growing landfill interrupts these processes with its own unique concoction of gases, liquids, and solids, which mix with (and travel well beyond) the landfill's immediate surroundings.[10]

Four Corners' employees and neighbors readily held the landfill accountable for bodily and environmental disruptions, either to subtly express disagreement with coworkers or bosses or as an explicit political tactic. Sometimes this tendency was playfully expanded upon, as when people would blame the landfill for bad cell phone reception or for changing the weather. But the environmental concerns that landfill workers and neighbors expressed were typically registered through affects that threatened to disrupt the body. Exemplary in this regard is disgust, the irruption of which calls into question landfilling's implicit care for bodily form. To avoid contamination from leakages and reconstruct bodily sameness, landfill workers clothe and bind their bodies in second skins.

These techniques offer a suggestive parallel with the efforts of local residents, who also seek means of representing and resisting perceived landfill leaks. Those who work and live near landfills separate themselves from their domestic waste, only to see the mass waste of other people and places accumulate before them. Yet it is also true that all of Four Corners' critics and protesters have been privileged to know places free of mass waste, just as many of its employees enjoy such freedom when their shift ends. Landfill owners and operators often resent opposition from nearby residents for this reason. They may accuse complainants of being NIMBYs. This pejorative term is derived from "not in my backyard," a label that is meant to characterize the motives of landfill opponents as selfish and ignorant. NIMBYs do not have a valid argument to make, according to this caricature—they simply live too close to something they do not understand and, therefore, fear. Four Corners' managerial team has sought to combat what they perceive as local ignorance by engaging and educating the public in different ways: at meetings to discuss their proposed expansion, they focused on clearing up misunderstandings and allaying fears; for a time they held annual open houses, with guided tours and information on display, including samples of the landfill's polysynthetic liner; they also welcomed visits from college students and researchers—including myself—to observe and document their operations.

But disliking landfills is not simply about the absence of knowledge; it reflects an awareness of very real and hazardous possibilities. If anything could be in the waste that goes into a landfill, then anything might come from that landfill, including contaminated air, soil, and groundwater. In this respect, the gulf between North Americans who do not know landfills and those who grow them for a living is not so wide. My coworkers at Four Corners knew the landfill intimately, but this did not prevent them from suffering discomfort or disgust in its presence. Expertise and knowledge matter, but whether someone is a waste worker, county regulator, aggrieved NIMBY neighbor, or visiting college student, they also inhabit a body that provides an anchor for their senses of place and pollution.

Landfills spread bodily harm *and* health in unseen ways, and these unacknowledged absences shape how people relate to waste subtraction and circulation. On one hand, North Americans have grown accustomed to sanitary transcendence of filth through mass waste removal, which amplifies their feelings of exposure and risk in the presence of landfills. On the other hand, achieving such transcendence tends to amplify the actual threats that both the subtraction and disposal of mass waste may pose to bodies. The experiences of those who work at or reside near waste sites demonstrate the unseen impacts of mass waste somewhere else on someone else, but also force us to consider the ways in which we all inhabit surroundings and bodies that are not wholly of our making, but are partly created by someone else.

LANDFILL BODIES

In some ways, landfills elude understanding in their very design. Incineration purifies, or appears to, but it may release a visible trace in the form of smoke. Burial, by contrast, hides. Early North American landfills did leak into neighboring locales, but this was imperceptible. Secretions of liquid and methane would collect and spread where they could, and only the occasional unexplained illness or exploded building nearby suggested that the serene surface concealed more disturbing mixtures below. To address growing concern about such invisible hazards, the United States passed sweeping environmental reforms through the Resource Conservation and Recovery Act (RCRA, pronounced "rick-raw"). By the time it came into full effect in the early nineties, RCRA had already brought massive changes to waste management. Using federal protocols as a guide, along with subsidiary state- and county-specific regulations, landfill owners must take elaborate precautions to avoid controversy and stay in business. If they are caught failing to adhere to standards for mass waste containment, they may be penalized or closed down. This means managing fires, vermin, and disease, as originally intended, but also emissions into the atmosphere and the release of fluid into underground aquifers.

Among my coworkers and in the technical literature on waste management and engineering, landfills that take these steps are frequently contrasted with mere dumps, which represent unthinking and unregulated disposal. At a typical North American landfill today, mass waste is entombed, cordoned off from its immediate surroundings in a state of suspended animation. Old-fashioned dumps allow waste to disperse unchecked through air, soil, and water. The modern sanitary landfill is more like a body. Its insides may be concealed, but it is provisioned with innards and orifices to allow for the circulation of fluids and gases. Landfill construction is a process of molding mass waste into shape, compacting and compressing it into taut muscular form, so that eventually it can be capped with a skin of soil and grass.

All standards are imperfectly enacted in practice. Similar to other construction operations, landfills require different types of workers ranked on the basis of skill. At Four Corners, most of the workers wear uniforms, work hourly, and are in unions. Operators are primarily employed to operate the multi-ton earthmoving machines, which tend to the growth of the landfill's body. At the time I worked at Four Corners, around a dozen operators were employed at any given time. The regular use of so many machines also makes it necessary to have a handful of in-house mechanics to fix and maintain them; when I worked at Four Corners, there were three. Both mechanics and operators were unionized, and only they were authorized to operate vehicles.

The employment of skilled machine operators and experts is important. Adding waste to a landfill is not as simple as filling a hole or piling more on top, like wet sand slapped on a sandcastle. Landfills are carefully and skillfully sculpted as they grow, a process that more closely resembles augmenting a living body with new implants. Using diggers, operators surgically peel back a patch of the landfill's skin in order to expose its flesh of mass waste. The exposed section may consist of a trench (for dumping sludge, asbestos, mobile homes, etc.) or an area, several acres wide, known as "open face." Patients on the operating table are covered in special garments that leave most areas unexposed while facilitating access to what lies beneath the skin; at Four Corners, autofluff was used in this way in dumping areas. Consisting of bits of rubber and plastic from recycled automobiles, autofluff provides a cushion, a tractable barrier to facilitate access of large, heavy vehicles to trenches and open face. Using bulldozers and compactors, operators gradually fill the hole with waste material and crush it into the existing slope. When the exposed cell or trench is filled, the skin of soil and grass is replaced.

During surgery, a patient's processing of fluids and gases is carefully monitored and, if necessary, manipulated to ensure a stable procedure and healthy recovery. Similarly, as flesh is added to the landfill, the existing network of pipes and pumps within must be carefully monitored. Leaks make sense only when there are presumed boundaries. As a consequence of RCRA, landfills require both a skin, to serve as a boundary that protects their flesh of mass waste, and an artificial network of organs—pipes to help them excrete and respire, liners to assist with incontinence. At Four Corners, the independent company that operated the gas plant employed a few more salaried workers. Together with the landfill employees, they were responsible for monitoring and assisting with the landfill's exhalation, digestion, and excretion. This work cannot be done through machines, but only by hand. It involves installing new gas wells, replacing leachate pumps, monitoring the gas field, and adding to the internal network of pipes as necessary.

Gas-plant technicians focus on the gas and leachate innards, and operators on implanting more flesh, but only the managerial team is focused on the growth of the landfill's body as a whole, planning its future expansion and setting the limits of its growth. The management team at Four Corners oversees the landfill's expansion, health, and fulfillment of regulatory standards. Although additional managerial staff were occasionally employed, during my time there the management team consisted, at minimum, of a manager responsible for allocating work to personnel and for day-to-day operations (Bob), his boss and the overall site supervisor (Big Daddy), and an environmental engineer responsible for meeting regulations, applying for permit approvals, and liaising with regulators (Corey). These are among the highest-ranking positions at landfills and tend to be salaried. Like the landfill's additional staff—including a few sales and administration personnel—they are not unionized, are more likely to have offices on site, and do not wear uniforms. With the exception of Corey, landfill management could be identified by the trucks they drove around the site. Most often they were parked at a distance, observing the entire landfill and its various operations. The landfill's entire body, not just the individual workers, was their responsibility. Big Daddy, Bob, and their associates sustained this global view through constant motion, taking in the landfill's whole body in an effort to keep it running according to their standards and those of regulators.

The standards of modern environmental regulation are upheld not by knowledge alone, but through cultivated senses. Managers would regularly discuss their perception of odors on site, where they smelled them, what they smelled like, and how strong they were. This was important in order to avoid complaints from neighbors, government inspections, and fines. Like managers, regulatory inspectors also rely on their bodily senses to achieve desired standards. They too must take in the whole landfill, though they do so far less often. Once again, the role of bodily comportment in regulatory practice is most obvious in the case of odor complaints, for which the noses of inspectors offer independent verification. Odor inspectors help translate odor complaints into admissible evidence in the governance of landfills. Odor problems fall under the county's Air Quality Management Ordinance, which is administered by the Michigan Department of Environmental Quality (MDEQ) and the County Department of the Environment (CDOE). According to the ordinance, landfills are noncompliant if odors are detected at level 2 or higher on an odor-intensity scale ranging from 0 to 3. Odor intensity standardizes measurements of poor air quality, in contrast to the subjective perceptions of complainants, who may differ according to their personal thresholds for a given odor or their past experiences with it. In addition to odor intensity, an inspector records a description of the character of the odor, its duration, and the weather conditions and time during which it occurred. Inspectors from the county or state actually receive special training that is meant to help them identify specific odor characters and quantifiable intensities in a reliable fashion. The result is sensory capacities that belong to no place. Tied to neither community nor corporate interests, and able to abstract landfill odor from surrounding environmental noise, inspectors are meant to assess actual odor impact.

With discriminating noses educated through the placeless resources of the state and the environmental science establishment, inspectors are empowered to translate the emplaced suffering of residents into noncompliance inspection reports, which, in turn, may lead to notices of violation (NOVs) and eventual punitive measures. If a landfill is found to consistently produce odors of 2 or higher, as has happened to Four Corners on more than one occasion, an NOV is submitted to the management team in writing, which stipulates a deadline by which they must correct the problem to avoid accumulating fees.

By testing landfill containment through the bodily metrics of inspectors, the MDEQ and CDOE also provide local residents and landfill employees with a means of contesting their claims. Timothy, a county inspector, was intensely disliked at the landfill and known for his interest in stray odor. Bob told me that one time he was forced to escort Timothy around the site for hours to locate a smell that the latter insisted emanated from the landfill. They searched throughout the property until Timothy located the source: a dog carcass in the woods. What is at stake in this story is embodied evidence of the landfill's realization of regulatory standards. The health of the landfill's virtual body—determined on the basis of its successful containment, continual growth, and profitability—is contingent on the way actual human bodies take it in and give it shape.

LIFE ON THE GROUND

Because of their overarching gaze, regulators and managers do not learn the landfill as intimately as the common laborers who make up the remaining employees at most disposal sites. Because laborers routinely work outdoors, exposed to their surroundings, their experiences place into relief the intimate relationship between bodies and places that most modern enclosures endeavor to conceal.

Work as a landfill laborer amplifies the everyday interanimation of body and place. Outdoor, low-status labor, in general, brings body–place mixtures into relief because workers are not enclosed from their surroundings (hence the class origins of the term "redneck"). After months of picking paper, I had lost almost twenty pounds, I was stronger, my skin had tanned after repeatedly burning, and my hands and feet had blistered and callused over and over again. For a time, exposure to the sun had so lightened my hair that family members thought I was dyeing it. Most North Americans are not exposed to their outdoor surroundings for such long durations.[11] If most of us wish to tan, change our hair color, or lose weight, this has to be scheduled into our

FIGURE 2. Laborer talking to management.

daily lives and will likely require that we use specialized equipment in private, indoors. Some of my fellow laborers had even darker and rougher skin than I, a material record of years of hard, outside labor. My coworker Eddy claimed that he had developed a hunch from picking garbage off the ground, which made it difficult for him to arch his back. Others blamed the landfill when they would get sick and miss work. It was generally accepted that laborers were as much shaped by the landfill as they shaped it.

Life on the ground gives laborers a different point of view. I routinely witnessed leaks of different kinds that the landfill was meant to contain. One could easily spot gases bubbling up at different spots around the landfill's base, evidence of methane that had not been captured by the gas field and was escaping into the atmosphere. I once tried to point this out to an inspector but was promptly rebuked by a landfill manager, who warned me never to bring up possible infractions if inspectors did not mention them first. Laborers typically lack union membership, salaries, and offices. Uniformed but uninformed, they are the least skilled and least well paid of all landfill employees. They are not qualified to officially recognize leaks.

But laborers do get to know landfills intimately. At Four Corners, my primary responsibility was picking paper. Papery materials are among the most common found in North American mass waste, which is why the name "paper picker" is commonly given to laborers who do more than pick paper and, as I will discuss below, also do more than pick. Being assigned to pick paper meant that I had to patrol areas where stray waste might collect, including perimeter fences, retention ponds, access roads, and the ridges encircling the landfill slope (known as "berms"). Stray garbage was landfill dandruff, sloughed off from its scalp and wandering freely in the open. While paper and plastic are not necessarily the most polluting materials, they are susceptible to being carried by the wind well beyond site boundaries, and this is a cause for concern. Loose garbage violates the ideal of landfill containment. In my first weeks and months, I picked almost exclusively and would spend hours filling clear plastic bags with materials (only some of it literal paper), leaving what I gathered along the road to be picked up and dumped back up top.

Like animal skins, landfill surfaces simultaneously contain and display—establishing a barrier between their insides and the surrounding environment, and demonstrating to onlookers that they have done so. Mega-landfills such as Four Corners are mountainous in size but designed to look like naturally occurring slopes. The productivity of such indistinction was a major factor in the rise of the landfill to national prominence after World War II. Laborers are responsible for keeping up these appearances. The type of stray waste we would find and collect varied, depending on where picking was done. Loose paper and plastic were more likely to be caught up on fences, berms, and ponds, whereas access roads more often collected bits of autofluff that had clung to the bottoms of trucks while they were dumping, or sometimes bags of trash they had accidentally dropped. The latter were more commonly found on the exit ramp that truck drivers used to descend the landfill and on the South Road they took to leave the site. I spent so much time picking the exit ramp and South Road in my first few months that some of the other laborers began calling this area "Joshua Street."

Picking up garbage would seem to be a relatively straightforward activity, but it was not always clear whether I was meeting expectations or maintaining the landfill's image as a contained body. When paired with others, I often felt as if our standards differed, especially concerning when an assigned task was completed. Landfill managers rarely settled the matter, because they appeared to see the landscape differently as well. This was partly due to their alternative mode of locomotion. Driving by in their vehicles, managers would occasionally mistake something for garbage that was not.

Despite these different perceptions of good picking, rarely was one openly criticized for picking poorly, except possibly on rare occasions when the landfill's body was on public display. In addition to government inspections, while I worked there, Four Corners hosted an open house in the summer. The site was closed to dumping, and laborers were asked to clock in beforehand to make sure it was ready to be toured. I had spent the previous evening picking the South Road, so when Big Daddy told another laborer and I that he had seen "twenty or thirty pieces," I was incredulous. As it turned out, most of the pieces he had identified from his truck were shiny leaves he mistook for paper, but which I picked and bagged anyhow.

The first time that I felt like my work was meaningfully evaluated was when it was assessed by a county inspector. I was picking on the exit ramp, which despite my efforts never seemed clean. By design, trucks gradually circled the landfill to ascend to the dumping area and exited on a steep incline. As they descended the slope, autofluff would spill over from up top, trickling down onto the gravel road and cascading into the ditches on either side. On a bad day, stray autofluff made it appear as if the whole top layer of the landfill was sliding too much to one side like a poorly adjusted toupee. I knew to pick autofluff and trash from the road itself, but I didn't know how high to venture up the slope, given the danger of swiftly exiting trucks, nor where the road officially ended and the autofluff toupee began, or whether autofluff cascading into the ditches on either side of the ramp was also a concern. But the visiting inspector complimented my work (a rarity, I would discover) and clarified for me where to pick and where not to pick. My field notes show what this meant to me at the time:

> [D]irty ramp, as dictated by . . . regulations, as interpreted by inspectors, equals fluff on the road (not on the side or spilling off down hills, all places I'd been picking). Sudden relief on my part—I know where to pick! Without direction I felt like the task was endless, and for that reason pointless. . . . [I now know] how to be evaluated and that my actions translate into actual procedures and codes and policies. . . . I am necessary!

Picking paper can feel like an endless and thankless task, but it is a crucial aspect of embodying a regulatory ideal. With inconsistent or illogical instruction on how to meet these standards, it can grow frustrating.

FIGURE 3. Perimeter fence in need of picking.

Frustration also arises because in some ways, laborers get to know the landfill better, given that they often have to move across its surfaces by foot. In their vehicles, visitors, bosses, operators, and more privileged laborers were shielded from the weather and had better access to toilets, shelter, and refreshments. Most laborers were alone to tend to their bodily needs. Relating to an environment through machines allows a person to skim the surface without participating more directly.[12] More immersive experiences of the landfill are worn on the body in the form of calluses, sunburns, cuts, aches, and pains of all kinds. Laborers use additional equipment so that they can roam the landfill in all temperatures and conditions. In the summer: sunscreen, a cap, and bottled water. In the winter: a hat, two pairs of gloves, two pairs of socks, long underwear, coveralls, and a thick winter jacket. Though hardly the air-conditioned cab of a vehicle, these vital instruments and second skins serve to mediate between worker bodies and filth or weather.

Laborers Eddy and Todd joked that their favorite time of year, "Christmas at the landfill," was when the bootmobile came in the summer and the landfill company paid for one new pair of boots of each worker's choosing. Eddy and Todd's evident glee was not completely in jest, since the wear on a laborer's boots is drastic after a full year. Operators and other workers are far less interested in bootmobile day; one even told me that he'd had the same pair for over five years but kept taking the new ones home anyway. Others bought boots that were impractical, some that I was explicitly warned by my coworkers not to get (steel-toe boots, I was told, freeze your feet in the cold). Laborers would have to constantly manage their outer coverings to feel comfortable and withstand a day's work. Such matters often took precedence over remaining clean or safe; sometimes being filthy might be preferable to bodily discomfort. While picking in the woods one winter I became so overheated, for example, that I removed my winter hat and the full-body snowsuit my coworker Timer had given me, and hung them on a tree like a scarecrow. Later that week, I learned that one of the managers had seen my scarecrow from their regular perch up top. He was so far away—and laborers are so iconically linked with their protective garments—that he had confused my second skin for me and assumed I had been standing around instead of working.

Even clothed, human bodies are uniquely exposed to their surroundings when compared to other animals, although both manage to transfer heat in similar ways (e.g., through movement and transpiration).[13] To transpire, animals need water. When I began working at Four Corners, there was little mention of the necessity of providing laborers with fluids. If we were lucky enough to be picking on the northern road, we could walk to Mac's house and drink from his hose, taking care to avoid Tina Dog, the lovable but protective mutt he kept chained up out back. A few months into my employment, the management team issued a new directive requiring employees to carry fluids with them on hot days. We were told, officially, that we should return to the maintenance building for water if we ran out, or ask people with vehicles to bring it to us. In practice, however, we were often forgotten about and would be presumed guilty of avoiding work if we returned too early before lunch for any reason. In response, Eddy suggested that we rebelliously work to rule— that is, drink as much as possible so that Big Daddy had to keep stocking up the fridge with beverages. Bathroom breaks were looked at similarly: officially they were acceptable, but we all knew that if we went too frequently we would be violating implicit rules of labor discipline. Instead, laborers were accustomed to relieving themselves wherever they worked on site. I was perturbed, on one occasion, when asked to cut down a row of trees and bushes at the base of the landfill that I considered my bathroom. Eric believed that if laborers had a union, we would begin by negotiating bathroom and water breaks and would get access to temperature-controlled cabs in the heat or cold.

Hydropolitics are also implicated in a landfill's release of fluids. Any liquids that cascade down a landfill's slopes or leak from its bowels are channeled into either retention ponds and ditches or underground pipes, ultimately to be collected within the site's designated leachate tanks. At many landfills, leachate tanks empty into sewers or process the pollutant for another use. At Four Corners, they empty into a water wagon, a tanker that hauls leachate up top to spray on the newly dumped garbage. This is meant both to stop leachate from entering groundwater and to improve microbial digestion within the landfill, to increase methane production.

Landfills are adorned with second skins too, as part of embodying the ideal of containment at the surface. At Four Corners, the final skin of soil and grass—the landfill's cap—is artificially grown on site. In conventional dumps, fluctuations in temperature could lead to fires. The evapotranspiration of a sanitary landfill's grassy skin also helps it sweat, in a sense, concealing its combustible guts underneath a more chemically balanced surface. The soil-and-grass skin comes from the compost heap on the edge of the site, where people dump yard waste for free and operators and managers see to it that it is cooked just right. Some of those same producers of yard waste return to collect the compost when it is finished, usually for horticultural purposes, also free of charge. Bob, who had always wanted to farm Michigan land like his elders, was proud of the soil's quality. After the composted green waste is applied to a capped section of the landfill, laborers manually distribute grass seeds with a spreader. The angle of slopes and the winds up top can make this difficult to do evenly across the surface, and seed is regularly spilled during the process. Over the course of the following year, the former farmers and construction workers in Four Corners' employ anxiously observe the progress of the landfill's only meaningful crop.

Concerns about the landfill's secondary, skin-like surfaces were a frequent source of discussion among laborers. Management wanted the grass to grow thick and even and would routinely send operators to mow and sculpt it into shape. When Ned was in Bob's favor, he would often mow. It is possible that his agricultural skills benefited him, because other laborers claimed that he did an impeccable job. When Ned and Bob began their public feud, another operator with a construction background took over. The laborers I spoke with disliked his landscaping style. Mac and Timer, who were brothers, may have been more invested in this because they too were responsible for tending to the landfill's second skin and, during the summer, were usually assigned to mow and weed-whack the edges of the site where larger machines could not reach. This was work they both enjoyed, compared with the endless travails of picking paper.

In truth, all laborers were invested in and beholden to the landfill's surfaces, which we wandered and explored beyond our official duties. Traces of our movements were left behind in the form of footprints, littered cigarette butts, and bodily waste. Some would occasionally leave more iconic inscriptions in the landscape, direct representations of *their* having been there. On the South Road, the name of a former laborer had been scrawled into wet concrete years ago. Eddy would constantly inscribe his name in the ground, as if tattooing the landfill's skin as he had frequently done to his own. Over the course of one week, I saw him write his name in dirt with a broom and then with muddy water dripping from a plastic bag he had pulled from a ditch. Signs of laborers were riddled across the landfill's body, and vice versa.

IDIOMS OF DISGUST

The ability of Four Corners to realize the twin ideals of containment and invisibility is reliant on the labor of landfill employees and regulators. Just as regulatory norms are imperfectly embodied, moreover, so too is the application of actual labor power attenuated by difficult environmental conditions and threats of contamination.

Within the collective formation known as "liberalism," sovereignty over one's own body is thought to be foundational to a good society. Without this, it would not be possible, for example, to sell one's labor to capitalist employers for a wage. But bodies do not always comply with the demands placed upon them. People get sick and they get tired. It may even be that they are not physically capable of withstanding the carnal reality of certain kinds of labor. Many of my coworkers at Four Corners struggled with the filthiness of their profession. In some cases, this meant that they tried to rise above their dirty occupation in their lives outside of the workplace, to buy things or provide for their children in order to transcend their circumstances and anticipate better futures, which I discuss in chapter 2. Some of this anxiety was expressed, also, in day-to-day work routines, through idioms of disgust.

The relationship between disgust, the body, and the person is a subject of considerable debate. Disgust is variously depicted as both the most moral and political of experiences as well as the most instinctual and reactive, the most metaphorical and the most fleshy and concrete. Most analysts begin with the assumption that our insides are actually or ideally divided from their outsides, which is premised on the ultimate isolation of each organism in a loosely contained body—disgust is therefore the psychophysical response of a subject whose virtual wholeness is threatened by holes.[14] Like all landfills, all bodies leak. If bodily immersion in an environment is understood as a threat to corporeal integrity first and foremost, then disgust makes sense as a symptom of this basic problem. Alternative models of the person–body relationship instead emphasize the continuity between bodies, other beings, and their external surroundings, especially through the exchange of material substance.[15] In this way, we might rethink the ideological sovereignty and sanctity of the individual person, as well as the paradigmatic image of disgust as the bodily whole threatened by its holes. Even in North America, there are many social situations in which the (possibly dangerous) sharing of substance overrules fears of contamination and bodily apertures become areas of exploration and enjoyment.

Rather than an inflexible instinct, disgust might more productively be imagined as a reliably produced *affect program*—that is, a reaction to environmental stimuli which, though automated, is differentially expressed and encouraged as an emergent outcome of ecological relations, bodily development, and social practices.[16] An affect program may momentarily overtake the self-conscious, social person but does not thereby owe its expression entirely to inherited predispositions. My coworkers and I tended to consider disgust an uncontrollable bodily affect that had to be triggered by an unpleasant experience, but this irresistible affect did not thereby manifest itself in uniform ways or at identical moments. This is because our encounters with troubling transience were triply mediated: by the general affect program of disgust, phylogenetically inherited and ontogenetically developed, and the singular life circumstances that brought each of us face-to-face with mass waste and each other. Sudden, disgusting mixtures of bodies with surroundings could therefore lead to nausea and fear or serve as a source of enjoyment and social identification, or both.

Second Skins

Like the body of a landfill, the body of a landfill worker has to be bounded and monitored to prevent unwanted mixtures with its surroundings. Regulatory practices attempt to discipline the isolation of the landfill's body from the surrounding area and its inhabitants. For a landfill worker, bodily exposure is self governed, specifically through the use of second skins: uniforms, boots, and gloves.

For most employees at Four Corners, every shift would begin in a similar fashion, with employees arriving and entering the locker room to change into their work clothes. Some came wearing their work attire already, but the majority of us would keep our work boots and uniforms on site. If there was a special duty to perform, there would be further steps to equip and adorn ourselves for the task ahead. As a laborer, I might be assigned to wade in the wetlands to pick loose trash, which required larger waders to pull over my regular boots. If I were asked to clean the vehicles, a full-body, plastic HAZMAT suit would be in order. The basic cotton uniform was a light gray, short-sleeved, button-up shirt with dark gray pants. Employees would acquire additional clothes to cover the uniform or to drape underneath it. Before I had a uniform and a locker, I had to launder my own clothes at home. It was worse getting my own clothes dirty and damaged, partly because I looked more like the un-uniformed managers, but mostly because I had to bring the odors, stains, and sweat home with me every day. After I got particularly filthy one day in my first few weeks, mechanics and operators laughed at me in the break room and said I looked like I had had my "ass kicked." People routinely equated my strange appearance to my lack of a proper uniform, perhaps hinting that I did not quite belong there.

Though uniforms cover most of a worker's body, gloves play arguably an even more important role by mediating the manipulation of offensive materials. After we had our clothes on, Bob would inaugurate the start of every workday by handing each individual their black, cotton work gloves in the break room and giving them instructions. This performative gesture emphasized work discipline in two senses: first, through Bob's control over our activities and his knowledge of where to find us on site and monitor our progress; second, through our own self-government as workers responsible for our own safety, beginning first of all with donning work-appropriate attire. Gloves were not readily discussed at the landfill in terms of disgust, but their value speaks volumes about the central importance of establishing barriers between skin and filth.

The soft, solid quality of gloves is meant to stand in stark contrast to the clinging fluidity of the most abhorrent substances one encounters at landfills. Laborers quickly learned the benefit of having more than the one pair of gloves we were given at the beginning of our shift. I was shown in the first week by a coworker where Bob kept the stash of gloves, so that I could take extras with me, for when my pair wore out or got wet. I was taught to pull a pair of latex gloves over the top of my cotton pair on wet days, protecting them from wet and slimy substances but also keeping them warmer on cold days. I remember the relief it gave me, knowing that there was soft material enveloping my hand and separating it from a piece of rubbish, a clump of sewage, or some unidentifiable grease clinging to a machine. I also remember the disappointment when I would notice a rip exposing my hand to these things, the sense of comfort at having another pair of gloves in my back pocket, or how appreciative coworkers would be if I gave them my extra pair when theirs wore out. I became so attached to our cotton gloves that I began taking fresh ones home with me. To this day, years later, I occasionally still find gloves stashed deep in the pockets of old coats.

The ability of gloves to protect the hands served as a measure of relative exposure and, in turn, disgust. The most disgusted I ever became at Four Corners was when some stray trash bags fell off the back of a dump truck, after the driver had accidentally left the hopper open while descending the exit ramp. "I was more repulsed than I ever recall being today," I wrote:

> The bags disintegrated to the touch and I had to pick through their contents, probably a few days old: old corncobs, a maxi pad wrapped in a toilet paper roll and other miscellaneous trash from the kitchen and bathroom. It smelled like rotted food and I breathed through my mouth. After finishing [and amusing passing drivers, one of whom made a disgusted face in what I took as a gesture of sympathy] the smell was still on my gloves and, to my intense disgust, my bare hands. Flies chased me wherever I went.

At the end of our shifts, we would throw away our gloves and wash our hands in the bathroom between the break room and the locker room, where two forms of soap were available. That day, I did not wait to wash my hands at the end of my shift. Instead, I discarded my gloves early and picked autofluff with bare hands, hoping that the dirt and dust would cover the stench on my fingers and palms, panicked that it would get on my phone or into my body somehow. I eventually walked to one of the portable toilets near the Roach Coach, where Bob's sister cooked and sold food from a food truck, and covered my hands with the dry gel located there.

Throwing gloves away at the end of the day signaled the end of our shift and the end of our need for protection. And yet gloves could fail; the smells and substances of the landfill could still follow you home. Different workers developed their own idiosyncratic approaches to cleanliness to supplement gloves and uniforms. Eddy would change his gloves constantly. He also claimed to never wear the same pair of socks twice, throwing them out at the end of every day. Todd would carry an all-purpose rag with him everywhere he went, though I never knew specifically what for. There were also workers at Four Corners who expressed very little concern about bringing contamination home. One of the laborers, whose home was nearby, would wear his boots and uniform home, in some ways reflecting his inability to escape the presence of the landfill, which was clearly visible from his front lawn. Others did not feel like they could escape it, no matter how far away they lived or how many outer coverings they shed before leaving.

Enjoying Filth

Doing dirty work does more than inspire avoidance and obsessive purification; it can also be a source of discursive and bodily creativity. This is so because our fleshy dealings with the world exceed the impossible ideal of pure, sealed-off bodies.

Eddy and I would routinely invent games while we were picking together, which helped pass the time before our shifts were over. On one occasion, we had to remove old tires from the top of the slope and decided to roll them down to the road to see if we could get them to bounce over the fence. On another day, we heaved dirt clods and rocks into the air and tried to catch them in our gloved hands, getting our uniforms stained in the process. After I remarked that the landfill would make for good sledding, Eddy told me he once rode an old "Elect So-and-So" sign down the northeast slope and into a ditch: "I was going at least 25!" We would do these things, knowing full well we were not only enjoying ourselves but contesting managerial discipline, playing with waste rather than picking it. But we also knew we were going to toss our gloves and shed our uniforms before we left for home.

There were times when what would otherwise be considered disgusting became more socially productive. The very vulnerability of apertures and the riskiness of encounters with filth were a source of humor and fun. One prominent source of masculine amusement associated with disgust came in the form of joking about homosexuality. At Four Corners, it was commonplace to impute homosexual behavior to others and oneself. This is a relatively common aspect of masculine sociability and gender hierarchy among different classes of workers in North America, though a related tendency to adopt racialized self-identifications with otherness (as in the expression "We're the landfill niggers," sometimes uttered by laborers) suggests a more general and ambivalent association of waste work with subaltern status. In a place where the threat of pollution surrounds physical contact with one's surroundings, intimate contact between male bodies, generally presumed by my coworkers to be even more polluting, served as powerful source material for jocular verbal banter and social identification. Like exposure to disgusting experiences, furthermore, sexual joking and profanity of all kinds usually did not follow workers home.[17]

If male homosexuality was productive as a sign of moral contagion—a threat to heteronormative, masculine personhood—then the greatest source of physical contamination at the landfill, everyone agreed, was the sludge pits. Some found it more than disgusting, even unsettling. Mac had had nightmares about the sludge pits ever since the death of the Canadian driver. As a consequence, he felt uneasy up top. I did not care for them either, yet I remember how much fun it was when Todd and I were instructed to dump bits of metal into the sludge pit that we had picked up around the landfill in order to protect vehicles. Todd and I stood on the back of a pickup and flung the metal rods into the trenches one after another. As we watched them land in the fluffy soup with a satisfying plop, we devised a game of sorts: our goal was to get them to land just right so that they would stand straight up before they sank, like flags. This consumed our focus until the last of the metal pieces had been jettisoned and I, for one, looked forward to doing it again. And yet, it was Todd who first warned me about sludge: "If you get *that much* on you," he had said, spreading his thumb and forefinger three inches apart, "you smell all day." This real risk of contamination and vile smell added to our enjoyment, precisely because it was something no one could handle or master. Gloved and booted, our extremities became zones of privileged exploration and risk, upon which we could focus our concern and care, to ensure that contamination did not threaten the whole person/body upon returning home.

Identification and Dirt

Not all workers at Four Corners were exposed to the same amounts of disgusting experiences. Growing a landfill involves transforming the surface of the Earth, which releases particulate matter to travel in the open air. This can be particularly difficult for those on foot. When not hauling leachate up top, the operator of the water wagon sprayed fresh water on site to reduce dust. Spraying water on the roads is meant to keep dust from migrating around and off site. Second skins are of little use when pollution is airborne. I learned this firsthand when I first joined the operators and truck drivers dumping waste at the top of the landfill:

> Most disgusting experience yet. When I was clearing shit off the flatbed into the open face. Two trucks pulled up on either side of me and squeezed garbage out using gravity and some kind of press. The one to my right kicked up a cloud of yellow dust—insulation—that covered me and got into my mouth. I used my bottle of water to rinse and then cast it aside into the garbage. The air smelled like kielbasa [polish sausage], no idea why—a stale, old, ripe odor that filled my body to my stomach. Wasn't close to puking, but I could imagine it. So I breathed through my mouth for a bit, still tasting kielbasa.

The divide between workers in vehicles and those on foot became clear through dust complaints. Passing traffic would routinely leave us in clouds of dust. Because this was often—though not always—directly proportionate to how fast the vehicle was moving, it was interpreted as an index of intended insult and injury. Some laborers would shout abuse at passing trucks or vehicles for "dusting" them. This was a particularly frequent complaint in the summertime and was even worse than landfill odors, according to some. The experience of disgust I felt, surrounded by a yellow, sweet-smelling cloud, was all the more disturbing because it was not bound to an earthly substance, but in the air itself.

An enjoyment of dirty work could become a resource to contest and mock class divisions. Many landfill employees would snicker at landfill managers and company reps who wore pristine work boots and drove sparkling new pickups on site. These visitors were not typically given gloves to wear by Bob, thereby revealing their aversion to getting their hands dirty as well as their relative superiority. He did not control or oversee their labor, as he did ours, and they had no need of special protection to mediate their immanent sociality with the landfill's leaky, slimy, unpredictable environment.

In general, laborers thought that other employees were more squeamish about getting dirty on site. One day, long into my tenure as an employee, I was asked to dig the tracks of a bulldozer. This was a task usually assigned to machine operators, who would use their small trench shovels to dig the dirt and debris from the massive tank-like tracks of their vehicles after a day of crushing garbage, sewage, and other assorted materials. On that particular day, an operator was in a hurry to leave and I was given the small shovel and instructed to dig. The tracks were stuffed with thick, spongy, black material mixed with dirt. I was up top and couldn't smell much of anything, surrounded by a variety of thick odors: the steaming garbage, the dust from truck traffic, the sludge pit, and the perfume sprayers. I had made a lot of progress after an hour or so, having climbed into the tracks to dig out the waste that had gathered inside. By that point, unidentifiable muck had splattered all over me. One of the other operators came over to check on my progress and was disgusted that I had gone to such lengths. "You got some by your eye. Do you know what that shit is?" I just smiled and wiped with my glove as he walked away disgusted. I assumed—from his reaction—that it must have been sewage sludge caked beneath the dozer. After I finished for the day, I stripped out of my uniform and boots, washed my hands and face, tossed my gloves. I would have been far less comfortable mixing with muck if I thought either it or the disdain of my colleagues could follow me home.

Moral Disgust

Not all operators show disgust like the one I encountered. While digging out tracks was marked as a particularly unpleasant activity, it could only become a story if it bore additional social significance. My story about digging out tracks was worth remembering because of its social circumstances: it was exceptional for a laborer to be asked to do this, and I had disgusted an operator.

Many scholars of disgust argue that because it is grounded in corporeal revulsion, it is the basis for other forms of contempt or resentment of a more moral and interpersonal nature. For example, when idioms of disgust collide and unwritten rules of bodily self-conduct are violated, the usually implicit power dynamics of sanitation come to the fore. Before I worked there, Big Daddy had the bathrooms alongside the maintenance building locked for his personal use and that of important guests. This became the subject of a number of complaints and jokes at his expense, particularly about his obsessive need for cleanliness and sense of privilege. Big Daddy's justification for this act, however, was that when the bathrooms had been available for general use, foreign-born, Canadian truck drivers had desecrated them. While I could not verify this story, from the details about the alleged incident I have ascertained from staff, the mess Big Daddy objected to may not have been a result of disrespect or vandalism. The cleaning techniques and alternative disgust idioms of someone more accustomed to squat toilets could also have been responsible. Either way, when faced with unfamiliar toileting practices,

Big Daddy had the power to repurpose the main bathroom and thus recompartmentalize the uses of space at the landfill.

At Four Corners, stories about people experiencing unusual levels of disgust were especially popular. Jerry, the most senior mechanic, was known for being somewhat squeamish about filth. He would sometimes hit golf balls onto the landfill from the maintenance building during his breaks, but when I tried to return one that I had found while picking, he refused it with a look of disapproval. Once, in the break room, he openly criticized operators who would leave garbage in their cabs when they brought their vehicles down to be worked on. In the middle of his rant, one of the operators mounted a defense ("There's not always time, sometimes . . .") but was immediately rebuked: "When they work at a dump, leave it at the dump!" For Jerry, the maintenance building was supposed to be separate from the landfill proper. People were aware of Jerry's attitude. If anyone was especially sensitive to contamination, it became known around the site. When operators and laborers worked together to repave sections of the road, everyone laughed at how disturbed George, a senior operator, would have been had he been assigned to work with us. The operators agreed that he would probably have "puked" if he had to smell asphalt fumes all day. Before I was aware of his reputation, I once accepted a ride from George on his way to the maintenance building, so we could both take our lunch breaks away from the landfill. "Enjoy your lunch," I said as I got out of the truck. He responded with an anxious smile, saying he couldn't "in a place like this . . . at a dump!" The term "dump" is marked in these instances—its use by Jerry and George indicates anxiety about the landfill's leakiness.

It was as if people with too much disgust, whether better-paid employees, uptight managers, or strange visitors, were too separate from the waste to really be part of our workplace. Disgusting people, on the other hand, were those who were too close to the site, who could not separate it from their personal lives. One story about condensate serves as an illustration. "Condensate" is the name for the liquid drops that cling to the insides of the pipes that help separate and channel leachate and gas out of the landfill and into the leachate tanks and biogas plants, respectively. When pumps malfunctioned or broke down, the result was constipation—a disruption of the flow of materials out of the landfill. This would involve a small group of different employees, usually gas-plant workers, but potentially including site managers and laborers as needed. Pumps were located deep within the leachate wells, and sometimes a man would have to be held at the waist while he reached within the large, stinking pit to hook the pump and drag it out. On one occasion, I was with a collection of gas workers and managers and the former mentioned a previous employee of the plant who had worked a lot on the leachate system. They laughed about his bushy beard, which always reeked of the condensate that he was exposed to, and laughed even harder at the thought of his wife kissing him when he got home.

Here the body exceeds attempts to discipline and domesticate it and becomes a historical record of activities and surroundings. A number of employees told me that their children or spouses would occasionally recoil, noticing the trace of the landfill they'd inadvertently brought home with them on their skin, hair, and clothes. Bart's daughters, for example, have repeatedly told him that he "smells like landfill" and shouldn't come anywhere near them until he showers. These are reminders that the *possible* stigma of waste work can manifest at home, even as relations at home and with one's family provide opportunities to transcend this stigma in various ways. Several operators, in particular, expressed a tension between their work and home lives, between middle-class aspirations and the wage labor of their days and nights, which I will return to in chapter 2. Many of us would leave our second skins behind, but we would also lose work-place idioms of disgust, the many social resources we had available to enjoy and laugh about what we smelled, touched, and felt. The separation between work and home is a North American ideal that I discuss in more detail in chapter 3. It is an illusion that can be difficult to maintain when phantom odors follow you home, where smells in your hair and filth on your skin are less likely to be laughing matters.

ODOR STORIES

I have argued that disciplining the body to avoid contamination while at Four Corners is analogous to the ways in which landfills are regulated: both are treated as sealed but leaky containers, an embodied whole with holes. But these two "bodies" are not only similarly imagined and enfleshed; they mutually constitute one another as well. To the extent that my coworkers could imagine escape from workplace pollution, it was not only because of their adornment and sociality, but because they sought to inhabit places detached from the landfill's orbit. In a sense, their homes became additional second skins to facilitate purification.[18]

By subtracting waste and making it invisible, landfilling makes possible places relatively free of transience. Hospital-like, North American dwellings enclose our bodies from unruly possibilities and appear to fade into the background, sterile and passive stages for action rather than active ingredients of it. This is a virtual pacification only, because places cannot cease interanimating those who dwell in them. For instance, one would expect that as experiences of chemically induced revulsion become increasingly rare—as a result of the general subtraction of transient matter from North American dwellings—general tolerance for strong odors would decrease and the strength of the disgust elicited by them increase. Landfilling thus constrains the emplacement of dwellings and the enfleshment of persons—their material becoming—which is distinct from the actual impacts that landfills may have on the backyards and bodies within their immediate orbit.

For those living close to its boundaries, the landfill's impact on emplacement could represent a significant challenge. From the beginning, Four Corners was the target of regular complaints, most of which were made by people living outside Harrison, in nearby Calvin. Their two primary concerns, familiar to many waste-management operations, were increased traffic and odor. To address these issues, the company that owned Four Corners prior to America Waste had installed storm windows and air conditioners in many of the houses directly surrounding the site, but not everyone was made this offer or was willing to accept it. By the time America Waste took control, only some complainants were residents who had lived in the area all their lives; others had recently purchased expensive homes in Calvin's newest subdivisions, only later to discover that they lived downwind from Four Corners and that this could affect the market and personal value of newly acquired properties. Eventually these complaints from new and old residents would lead to a lawsuit and an out-of-court settlement, but during my employment the landfill's management team was still pursuing measures to address local discontent. Shortly before I began working there, they erected a sign in Calvin advertising Four Corners' proximity, with the specific intention of informing and forewarning potential homebuyers. Not long after, they received new complaints from realtors who were concerned that revealing the presence of a landfill would scare off potential clients.

Contaminated Atmospheres

For local residents, the landfill's impact on their bodies is frequently contested on the grounds of its transformation of local atmospheres, signaled most obviously by the production of odor. Although not usually thought of as a material ingredient of place, atmosphere is fundamental to any setting where breathing organisms might dwell. When basic atmospheric conditions for life are lacking, it radically affects emplacement, or the mutual constitution of body and place.[19]

The atmospheres that surround us are not an absence, but a positive medium through which we move and are enabled to live. I was once asked to clean out the water wagon when it accidentally tipped on its side up top and became filled with trash. Concerned that the sprayer might become jammed, Bob sent me to climb on top of the wagon's tank and lower myself inside. The tank was divided into a series of chambers winding around the inside, and just big enough to climb through one at a time, like a tiny submarine. I was initially worried it might smell, and relieved to discover that it did not. I had never received confined-space training, however, and had not been warned that my oxygen would run out if I stayed inside too long picking. Bob may have known that this was a concern, because he told me someone would remain within shouting distance while I was inside (though when I reemerged, no one was in sight). Not knowing any better, I picked for nearly an hour without coming up for air. There was not much trash inside, but I had to take regular breaks because I felt increasingly lightheaded and tired. It was only later that my wife, an environmental engineer familiar with confined spaces, informed me that I had been slowly suffocating.

Landfill odors disrupt the ideal process of middle-class emplacement that is supposed to set the stage for actions and events free of transient matter. Calvin Township, which includes the small villages of Brandes and Eatonville directly northeast of Four Corners, in the path of the prevailing wind, has been exposed to odor problems since the landfill first began operation in neighboring Harrison in 1991. Dust, plastic bags, and construction noise are all carried in their direction, but no leak is as contentious as the smells of rotting garbage, sewage sludge, smoldering compost, and landfill gas. Odor has plagued both Brandes residents and landfill managers since Four Corners opened; official complaints to the county and state environmental protection divisions increased dramatically in 2003 and 2004. Most residents tended to agree that landfill odors became much worse once Canadian garbage began coming to the landfill in the spring of 2001.

Maude believed that problems with Four Corners began to multiply when ownership of the landfill was transferred from a local Michigan company to one of the big three transnational waste corporations. The new ownership, coupled with new Canadian trash imports, created a different landfill: "The traffic got worse, the smells got worse. [Before that] it was more rare, it was like once in a great while, but if you called the dump directly and said 'man there's a smell over here' immediately they did something about it." According to Maude, the landfill used to be more sensitive to community concerns. Indeed, it was integrated into the community of Brandes in a further sense because, at that time, employees of Four Corners came to the store on a regular basis and she could recognize some of them as locals similar to herself. Now more landfill business came to Lions Service from tractor-trailers that broke down on the way to and from the landfill, but she remembered not long ago when her encounters with landfill employees were more personal and the occasional stench was far less of a concern.

When I was employed there, it was difficult for Four Corners to control its odor problems. In part, this was due to the fact that the managerial team was committed to growing their own compost and producing methane for the power plants on site. Here the corporeal metaphor I have been using to describe the landfill becomes more metonymic. While a landfill only resembles a living body, it is directly associated with and contains life, including many millions of microbial beings that reside within its bowels. These microbes, part of the kingdom Archaea (distinct from Eukarya and Prokarya), can withstand unusually high temperatures and pressures. The methanogenic archaea that subsist in landfills are similar to those in the human gut: they feast on the energy released from degrading hydrocarbon bonds and, in turn, exhale CH_4, or methane biogas. Four Corners is a "bioreactor landfill": materials are selectively added to it in order to improve this microbial digestion, which both increases the production of biogas for the plant and, by hastening decomposition, makes space for more waste and optimizes the landfill's profitability before its eventual closure.

The landfill's sludge contracts with wastewater treatment facilities in Detroit and Toronto were the source of the most offensive odors and what eventually precipitated local protests (see chapter 5). Some managers were convinced that the sludge was not worth the money, given how much was invested in eliminating odors, but it did provide excellent feedstock for the landfill's gut, a highly bioreactive appetizer that could kick-start microbial digestion.[20] It is not a simple thing to harness a landfill's digestion and respiration processes, however. The most favored laborers would often be appointed the task of maintaining the gas field alongside technicians from the gas-plant company, and during my tenure at Four Corners this was Todd. They would raise wells and monitor the gas field, expanding the landfill's plastic intestines as it grew and addressing the regular problems that arose. As the landfill grows, tall pipes sticking out of the side are swallowed up, surrounded by newly compacted flesh, and need to be extended so that the gas-collection network can grow outward. Gas-plant technicians and landfill employees would try to leave signs behind to signal to operators where the unraised gas wells were.

All of this work was part of managing the production of gas and capturing it before it could enter the atmosphere. Before waste was successfully entombed, however, the odors of bacterial decay could easily travel downwind. It was also the responsibility of Todd and landfill management to manage these atmospheric externalities, and they did so by burning wood behind Timer's house, much to his displeasure, as well as with the use of SEs (stink eliminators). The latter were perfume sprayers hooked up to portable plastic containers and fixed to cords that were suspended like telephone lines along the roads and up top. An invention of which Bob and Corey were very proud, an SE would be positioned so that the perfume would mask the scent of odor that accompanied microbial digestion.

Olfactory Emplacement

Most residents were unfamiliar with the microbiological and technical sources of their odor problems or the landfill's attempts at remediation. At Lions Service, instead, it was the corporeal effects of odor that mattered. Although many things were discussed as I sat ensconced each day in the coffee klatch at Brandes's main party store, conversation frequently turned to whether anyone had smelled the landfill that day and how bad it was. At the time of my research, several times a week, between 5:30 and 6:00 in the morning, the smell of rotting garbage followed by the stench of sewage sludge was common. Maude correctly attributed this regular succession of smells to the landfill operators digging new trenches (a process that stirs up old waste in the midst of decomposition), into which they dump the daily sludge loads. The large compost piles, assembled on the western side of the landfill for landscaping purposes, also produced fumes; when compost piles are regularly stirred to add oxygen to the batch of rotting plant material, they release a noxious odor.

Through conversations about the Four Corners, local residents worked together to document its disruption of a shared sense of place. On an informal basis, the odor talk of residents served to fix a shared orientation to the landfill, contesting the official narrative of its contained and bounded embodiment by spreading discourse about leaks. In addition to talking about the odor, residents relied on the MDEQ and CDOE to ensure that such infractions were independently recorded and assessed and attempt to have the landfill punished. In both cases, odor talk and odor assessment, the official landfill narrative is contested through a bodily basis of knowing place through smell.[21]

In Brandes, sometimes it can seem as if strange smells follow you wherever you go. According to Jerry, Gwen's brother-in-law and a tow-truck operator for her business, "On a bad-smellin' day, you smell it everywhere. Sometimes it's so bad, you think it's somethin' you run over!" As drivers for the towing service move up and down the main highway on a regular basis, they are able to report back on how far the odor from the landfill travels, as do customers from farther north in Eatonville. Once circulating in the space of the store, these accounts not only relay what conditions are like outside but also bear witness to an alternative cartography of landfill leaks, inscribed in afflicted bodies. At their most extreme, tales of odor encounters become performances of subversive articulations of body and place, as in the often-used phrases "I thought I was going to puke" and "It was so bad, I almost threw up." One could argue that the pleonastic pronoun *it,* so often used in English to refer to ambient environmental conditions (e.g., "It's raining"), helps to powerfully enact emplacement.[22] At Lions Service, people talked about smells so strong that they made it difficult to get up and go to work in the morning or made it necessary to close windows on hot days. Maude claimed that this hurt walk-in business: "As far as them coming into the store, if *it smells bad* I wouldn't come in. I would go down the street and get to another store before I would get out of my car. . . . Who has an appetite when *it smells like that?* I mean they're not gonna come in and buy food!" Strong odor is here registered as a bodily invasion, a virus, as if the place itself were turning against the bodies of its inhabitants. In exchanges about odor at Lions Service and the surrounding town, the afflicted body created a shared framework for talking about and knowing place. Each body stood as evidence of landfill leakage into Brandes and beyond.

In addition to evoking the corruption of bodies, the odor talk people shared expressed a violation of home, a challenge to the kinds of freedom from impurity that many North American moderns take for granted. Brandes residents described having to stay inside on beautiful days rather than have cookouts or play with their children. One woman conveyed her frustration and embarrassment at not being able to have an outdoor high school graduation party for her child. Landfill odors are thought to displace the fresh country air that attracts property builders and homeowners to the area and which long-term residents recall with fondness. The growth and maintenance of the ideal home are short-circuited in a way that is both familial and economic.

Sanitize and Quarantine

Atmosphere is central for dwelling, and smell for establishing meaningful connections to place. As a result, odors can be highly disruptive to ideal body–place relations. The absence of signs of decay and contamination that North American moderns desire and demand helps to make possible the presence of a more limited range of alternative experiences and memories. If some backyards are seen as corrupted, consequently, it is because landfills are taking from some that which they give to many others.

To be modern once meant living in a world of inescapable public health concerns and offensive experiences. During the rapid industrialization of the nineteenth century, city dwellers had to contend with a wide variety of nuisances and forms of pollution. More than just offensive smells, this pervasive quality of *noisomeness* in modern life forced people to reflect upon and reclassify their bodily immersion in the world. Lofty projects of social and corporeal reform like sanitary engineering were established to meet the challenges presented by industrialization and urban dwelling. The subtraction of waste and its disposal elsewhere played an important role in these initiatives, but they operated on a relatively simple and very old medical principle—that of quarantine.[23] Noisomeness and miasmas have not left modern life entirely; they are displaced elsewhere and kept secure, their impact reduced. Now fewer people are exposed to the waste of many. The effects of decay and disposal, once meted out through lived social space, have been moved into the countryside, deep underground, and far into the atmosphere. Landfills represent a dissimulating form of social relating, one where microbes, leachate, and sludge figure prominently, having displaced more familiar figures of social topology such as class, the built environment, and the body. But landfilling is a social technology as well as an environmental one: it redistributes and concentrates exposure to mass waste. As waste is unevenly distributed, so too are affective reactions like disgust.

For some of my informants, contaminated bodies and foul atmospheres remain a threat to bodily form. This brings into relief the immersive fit of bodies and places that many take for granted, but which the people I met were unable to. The odors coming from Four Corners ultimately led to a successful class action suit on the part of local residents, new and old. Despite the wishes of Maude and her friends and family, the law firm was not able to establish a claim on the basis of health effects. They were convinced that it was causing acid reflux in the community's children, and possibly even cancers, but did not have the expertise or the legal support to advance their claim in public. In fact, if the hygiene hypothesis proves correct, it might be the expectation of sterile, empty surroundings that is to blame, insofar as this reduces exposure to the helpful microorganisms that dwell in the filth we quickly eliminate. Maude and her friends may be right: landfills may be, in a way, responsible for apparent increases in incidents of acid reflux and other health conditions, not due to their proximity to a select few, but because they so efficiently subtract transient materials from all our lives and, in doing so, constrain how we relate to our surroundings and how our bodies develop.

This may be so, but research on the impact of hygiene on our health is ongoing and controversial. My own investigation failed to explore these possible causal connections, and so did that of their lawyers. Rather than focus on the role of contaminated bodies, the legal team premised the lawsuit on the landfill's impairment of their ability to enjoy their property. While successful, this relies on a conception of dwelling that obscures the mutual constitution of material surroundings and the bodies that inhabit them. Instead, the homeowner is conceived as a singular entity, attached to a commodity that they have a legal right to enjoy, free from odor. The ideal singularity of that homeowner is challenged by landfill leaks that afflict their dwelling but is performatively reasserted by compensation for their suffering. It was a condition of being rewarded under the lawsuit that one be a proven complainant attached to a household in the vicinity of the landfill. What is lost in this form of redress is an appreciation for how bodies and places mix and interanimate one another, an ongoing process that shapes the lives of those within the orbit of landfills and those far beyond them.

LANDFILLS AS BOTH SYMPTOM AND CURE

The body of a landfill is never perfectly assembled in accordance with regulatory ideals. Like an animal body it will inevitably leak, a little or a lot, into its surroundings. And this is where the corporeal metaphor reaches its limit.

Landfills are not bodies, because bodies die, decay, and eventually merge with their surroundings. The growth of a landfill has no real limit, save the impact it might have on other beings. Like many of its kind, Four Corners could not get much taller, for fear that flocks of scavenging seagulls might interfere with air traffic. It could not get much wider because it was bounded by a wilderness preserve on two sides.

Landfill workers and neighbors must manage actual and possible landfill leakages while attempting to make their bodies and lives hang together as they have come to expect. To the extent that this expectation is due to the subtraction of their waste, landfilling is the cause of their difficulties, the solution to them, and the reason both appear as such to begin with. By helping to maintain modern dwellings, possessions, and selves in a state of virtual, reconstructed sameness, the removal of waste constrains our immersion in places and relationships with materials. This fills up landfills and forms people resistant to their proximity. Without the expectation that everyone deserves to be free of transience and filth, mass waste could be seen as an invitation to recover and recreate value.

The tactics my informants pursued to cope with exposure to mass waste are no different, in principle, from the everyday lives of people outside a landfill's orbit. The liminal spaces of bathrooms and locker rooms facilitate bodily transitions equally well whether located at a landfill or a public gym, and people of all sorts adorn and compartmentalize themselves to engage with the transient materials that come from their bodies and possessions. Second skins can be found at all stages of the process, moreover. Folded into the experience of being at home are countless, seemingly banal decisions about what to do with troublesome materials. Toilets, garbage disposals, Dumpsters, and recycle bins are mini-landfills in our midst, a scaled-down version of the containment and invisibility that regularly occur in places like Four Corners. What so perturbs the landfill's employees and neighbors is that they have to engage in further compartmentalization. In this sense, the bodily experiences of workers and residents are only an amplified version of ordinary experience. The exceptional experiences of a landfill laborer or neighbor are only momentary, after all. The powers of subtraction have too successfully emptied most of their lives of undesirable experiences for even the most squeamish worker or nauseous local to be permanently aggrieved, even when bodies and places cease to articulate in perfect harmony, when bodies bring contaminants home, or when the home itself is threatened by unpleasant atmospheres. If landfill leaks readily elicit anxiety from workers and garner complaints from local residents, this is partly because the efficient subtraction of mass waste has raised North American standards, elevated what we consider acceptable surroundings.

In saying this, I do not wish to downplay the negative experiences of people who work with or live near sites like Four Corners. The landfill can be disruptive, and the possible health impacts are as yet unclear. My intention is to challenge the tendency to overemphasize the differences between people close to waste sites and those far from them, to highlight the social relations that connect most North Americans intimately, but unknowingly, to these hated sites. Some people are more directly and transparently affected by waste management, but this can happen to nearly anyone. Not many will work directly with or live close to mass waste in their lifetimes, but waste subtraction can and does break down. Occasionally sewage pipes burst and basements flood with effluent, or garbage collection halts and Dumpsters overfill; there are garbage strikes, neglectful municipal governments, and unregulated or criminal monopolies. In these moments, bodies are exposed and forced to mix, however temporarily, with untransferred transience. Waste management can no longer be forgotten, and we become surrounded by what our hands have flushed and tossed, by garbage that keeps coming. But such exceptional occurrences merely lend support to the general rule: for many North Americans, transience ends up elsewhere. This inevitably shapes what we think of landfills but also, and less obviously, how our bodies take form and last.

In this way, humble domestic acts of sanitary containment and separation partake of prevailing background assumptions about form and process. The reconstructed sameness of our everyday surroundings that landfilling makes possible has replaced noisomeness and transience with disposability and permanence—these are the virtual qualities we now truck with in our ordinary lives. The extraordinary experiences of landfill workers and nearby residents complicate this process of substitution. In so doing, they draw attention to the actual relations with transience that provide the material foundation for all our virtual lives.

TWO

Smells Like Money

IN DIFFERENT PARTS OF THE WORLD, the paradoxical profitability of waste work is encapsulated in an iterable quip indicating that waste is wealth. If an unpleasant odor attracted our attention at Four Corners, for example, someone might inhale deeply and respond, smiling, "Smells like money!"[1] Few idioms were repeated and circulated as widely around the landfill:

> At the Roach Coach, Tanya is preparing lunch. While we wait she explains to an audience of truck drivers and laborers that the new male employee she is training "acted like he was gonna puke today." One of the operators immediately responds, "You get used to it pretty quick, it begins to smell like money."

> While we pick paper off of the side road, Eddy yells at passing trucks, "Wooeee, I love that smell!"

On the one hand, this aphorism indexes a social divide between those who desire separation from waste and those willing to profit from this desire. The one who smells money reminds the rest of us what it is that makes waste workers different from waste makers—we benefit from their inability to see or smell anything but waste. The expression depicts the speaker as someone able to accept or even enjoy dirty work, whether or not they really can and do. By implication, those who complain or act like they might get sick can appear to be less savvy, less in control of their bodies, less masculine.[2]

In North American life generally, money serves as a gloss for the many things that motivate people, even those things that money cannot buy. But there are nonfinancial reasons not to smell the waste, too. The actual income that operators earn pushing and burying garbage is typically less than they would get from working in the construction industry; it is far more secure, however—if you lose your job at a landfill, it will not be because people stop wanting to get rid of their waste. For many laborers, working outside all day with relative autonomy is more desirable than an unskilled position in retail. Why wouldn't anyone work at a landfill?

On the other hand, "smells like money" indicates a more iconic, corporeal division within the speakers themselves. As shown in chapter 1, experiences of bodily disgust and contamination are real obstacles to working and living with waste. To assemble a landfill, one must find ways to transcend such corporeal limitations, and "smells like money" owes its popularity to widespread recognition of this fact. A simile that links waste to money is amusing because rotting waste is an odor my coworkers recognize all too well as something *other than the smell of money,* something radically different, in fact, that can linger on their person as a source of embarrassment and shame. This stigma can act as an indelible mark that affects not only a worker's body, but their happiness and sense of self-worth as well. Why would anyone work at a landfill?

This chapter explores different limitations and paradoxes of North American class and value that my former coworkers sought to overcome. Some of these difficulties are unique to their line of work, associated with the injuries and indignities of dirty and dangerous labor, but others are all too familiar. For my coworkers, and for many North Americans, *amounting to* an ideal kind of person is symbolically and causally tethered to *amounting* sums of money. The desirability of money comes from wanting pure possibility without limits, to transcend constraints. "Smells like money" is funny not only because waste smells nothing like money, but because the smell of money is completely irrelevant to its social function as a store and measure of value. Interpreted further, "smells like money" thus draws attention both to the worker divided and alienated by wage labor and to the immateriality of money itself. This is the strange quality that allows money to stand for almost anything, a virtual potency that belies its material instantiation in coin, paper, plastic, or digital code. Waste, by contrast, stands for the unpleasant excess of materiality as such, the inevitability that things will break down and threaten us with their impermanence.

As we have seen, work with mass waste threatens the body with contamination, but it also can threaten the moral and social figure of the person. This is so because it is presumed that people end up doing waste work because they could not amount to something better. My coworkers attempted to overcome these perceived limitations, and the stigma thereof, through representations of transcendent value. Their efforts to do so highlight the paradoxes of middle-class aspirations in contemporary North America that were coming to light even before the burst of the housing bubble and the start of the Great Recession. Transcendence has long been associated with American dreams of capitalist success, beginning with Max Weber's demonstration of an elective affinity between the accumulation of capital and the Protestant ideal of overcoming material attachments in pursuit of otherworldly rewards. The obverse of such self-denial is, arguably, investment in an individualist fantasy of self-creation. North American middle-classness is an expression of this fantasy. More an ideal than an economic reality, middle-classness is something people aspire to despite the real precarity of their circumstances. Through their labor, landfill workers help make possible the would-be transcendence and self-invention of North Americans: all that which might challenge a pursuit of permanence is removed from sight and dealt with somewhere else, where it can pose no barrier to projects of possessive individualism.

First, I will discuss how landfills translate collective desire for separation from waste into profits. I will show, in particular, how the class hierarchy at any landfill, which divides laborers from operators, mechanics, managers, and office personnel, is also a product of this wealth-accumulating process. With the late-twentieth-century shift toward neoliberal capitalism—a homegrown ideology originating, like Fordism, in the American Midwest—waste collection and disposal activities in North America were increasingly privatized and acquired by a handful of corporations. The relatively flexible work conditions that resulted have tended to grant workers more responsibility for governing their own labor power, especially in terms of safety. Following the work of David Pedersen (2013), I characterize "value" as an interpretive tendency of capitalist social relations, which have transformed wages into more than money—into a source of wealth generation, a means to finance homes, vacations, and college degrees. For many dirty workers, the future earnings of their children and the future equity of their most valued possessions serve as powerful signs of class advancement. But investment in these immaterial values is a risky proposition. Landfills may eventually prove more secure than investment in houses, cars, or college degrees, those storied stores of value typically sought by the aspirational middle class. If you can smell the money, that is.

Moreover, there are further limits to such mobility. The ability to realize middle-class aspirations has always been historically gendered and racialized in North America and remains so today. A person may need to overcome personal disgust and social stigma to work at Four Corners, but a woman or person of color would first have to be offered the privilege of doing so. Some bodies are not so easily forgotten; ideologically they are too situated, their specific identity too pronounced to be considered for the unmarked, abstract labor power of smelling only money. During my time at the landfill, the only women on site worked in sales, food service, or administration, and all employees were white. I learned that women had worked as laborers in the past but had experienced severe harassment and, presumably for this reason, had eventually quit. It is possible that managers avoided hiring female laborers to avoid similar problems, but I could not confirm this. Whatever the reasons, this had the consequence of creating a highly gendered working environment dominated by relations between white men. This certainly shaped my coworkers' actions on site when I was there, and likely amplified their already gendered experiences of class.

Therefore, the analysis that follows does not represent intersections of race, class, and gender in all North American settings, workplaces, or even landfills. Rather, it brings into relief some of the troubling ways in which work, value, and person can be woven together in North American class formations. For the white men employed by Four Corners, internal sources of stigma further divide workers from one another. Fulfilling the patrimonial ideal of supporting one's family can also mean supplication and humiliation while at work. Consequently, there are those who refuse the economic opportunity to advance in rank at the landfill. They do so not in defiance of the promise of transcendence, but as an alternative realization of it. Whereas some at Four Corners seek transcendence through class struggle, others see the social relations of work as a way of guaranteeing individual freedom. All attempt to earn a paycheck while avoiding shame and disappointment at work and at home, even as circumstances conspire to make these seemingly incompatible goals.

LANDFILL CAPITAL

A landfill is a congealed product of human activity and imagination, heterogeneous materials, and large earthmoving machines. As a product of these combined sociomaterial forces and relations, any particular landfill is imperfectly realized, may leak, and may smell. As an investment and corporate asset, however, a landfill is, in principle, like any other capitalist enterprise—it is a money-making venture first and foremost and is typically represented as such. Like any act of representation, seeing (or smelling) landfills as money-making machines is a partial and imperfect account. It tends to obscure or downplay the many other forms of social activity that go on at places like Four Corners, including acts of play, aggression, flirtation, friendship, accident, and wonder, which entangle everyday work routines with motivations and tensions that cannot be reduced to the kind of limited explanations typically glossed as "economic."

It is sometimes said or assumed that waste is a commodity for the waste retrieval and disposal industry. In point of fact, it is not garbage that is bought and sold and that dominates the lives of workers, but waste services, priced within a market of similar transactions. To understand how landfills turn waste loads into money, therefore, one has to appreciate how waste services are commodified, or bought and sold, which is distinct from how waste gains or loses value. At Four Corners, when members of the sales office are attempting to sign new clients, the particular contents of the waste loads are of no more concern than the particular qualities of the landfill workers (whether or not they are happy, funny, loved, confident). Both are represented in a highly abstract way. The waste does not have value, per se, but serves only as a raw material for the application of the landfill's abstract capacity (or labor power) to do the work necessary for its disposal.

Value in capitalism is a powerful and persuasive story, a systemic representational tendency.[3] During a typical act of capitalist market exchange, for example, a commodity reveals very little of all that went into its production, except the very generic quality of its having been made by others, which is all there is to go on when we tell further stories about what it costs, or its money-price. That story of value nevertheless changes over time as it is further reflected upon and reinterpreted or, as Pedersen (2013) puts it, transvalued. At Four Corners, money is not represented only as something to pursue, or for the sake of which workers endure filth and odor. Many of my coworkers may have been fond of the expression "smells like money" while on site, but they were equally familiar with sayings like "Money is the root of all evil" or "Money isn't everything" and could be called upon to recite them if appropriate to the situation.

At landfills, this interpretive process of commodification begins with the categorization of different forms of waste. The RCRA, introduced in chapter 1, distinguishes landfills on the basis of the waste streams they are allowed to receive. As a Type II landfill, Four Corners can receive municipal solid waste, demolition debris, contaminated soil, sludge, yard waste, and incinerator ash. Because hazardous waste is conditionally forbidden, the accurate classification of waste loads prevents fines from state regulatory agencies and provides a necessary paper trail to substantiate the continued legitimacy of the site. According to national regulations, waste-generating clients must produce a document known as a "waste profile" that verifies the contents and characteristics of waste streams, on the basis of which a sales representative can create a binding contract. Each label assigns a mixed load of waste to a particular class, with predictable environmental impacts and handling requirements. By rationalizing and mechanizing waste practices so that they are fixated around the possibilities of pollution, landfill services become valorized into a profitable form of business. Because a waste profile must exist before a contract with a prospective customer is finalized, it may involve formal distinctions that are impossible to observe in practice. The indeterminacy of mass waste—the fear and promise that anything could be in the waste—is attenuated by these classificatory measures, but only partially. It is impractical to inspect every incoming truck.

It is through the medium of contracts that waste services can be bought and sold on a transnational market as a commodity, because they reliably prevent waste from reentering the everyday lives of most people. Assigning different objects to general categories like "hazardous" or "ash" reduces the heterogeneity and individuality of waste. This is carried out, in practice, by way of a further reduction from abstract category to aggregate quality. After being assigned to a formal type, incoming loads are reduced to weight or volume so that a price or tipping fee can be assessed, which stands in for the cost of assuming practical stewardship of waste. At its most abstract, this is the labor power required to assume the burden of its negative value. Waste loads may be assigned a fixed money-price at the official signing of the contract between waste generator and landfill, for example stipulating a certain number of loads per day of a specified size. Or, as is also common, the weight or size of a given load may be determined during entry into the facility at the scale house, where additional documentation and measurement are required.

Attending to particular qualities of things and ignoring others is a common interpretive practice. In the case of landfill operations, however, this is performed more systematically, as the selection of a few aggregate characteristics establishes an interpretive frame by which all waste loads seem commensurate as exchangeable negative value. Establishing this base level of equivalence makes possible other forms of calculation in turn. Quantifying the incoming waste gives the landfill company a sense of how quickly the site is filling up and how much of its capacity or air space remains available. Many waste firms use air space as a measure of the value of their landfills and to assess their remaining years of profitability. In doing so, they abstract from the complex legal, environmental, and social conditions that make this space one company's private property and fill it with waste. Such conditions are, instead, subsumed within value as a dominant interpretive tendency. Actual things and actual people are not entirely missing from these calculations, but they are represented only in the most general way possible. Waste is reduced to abstract physical qualities (the weight, volume, and character of a waste profile). People are reduced to abstract labor power (the amount of time generally necessary for employees to bury waste), and landfills are able to turn a profit by paying less for this time than they earn by stuffing the greatest amount of waste into the smallest amount of air space in the least time.

Wage at landfills is a product of rank, whether one is a manager or engineer, a unionized operator or mechanic, a member of the sales or human resources staff, or a laborer. This divides landfill workers on the basis of their credentials and relative skill, which determines their relative class position. Management and environmental and sales specialists receive upper-middle-class salaries ranging from $75,000 to $100,000 annually. They are most likely to receive promotion to positions of greater authority and power within the landfill company and have the greatest job security. Though I have lumped them together here, the environmental and sales personnel hold more specialized positions and require college degrees, whereas managers may not (e.g., Bob began as a laborer with no college degree, as he is fond of telling people). A difference in credentials is typically reflected in earned income: though Bob had more responsibility, seniority, and authority than Corey, he made considerably less. It is also worth noting that from the very beginning of Four Corners' existence, these positions have all been filled by men, though women have occasionally applied.

Except for the professional and managerial workers, every position at Four Corners involves middle-aged men with seniority paired with new, younger employees in training, usually in their early to mid-twenties. Because advancement and earning power are connected, the result is an internal labor market governed within the firm itself. Seniority is typically reflected in one's hourly wage. Because of these veteran–trainee relationships, one need not have had prior training before beginning in any of these positions and, consequently, upward mobility between different jobs is not uncommon. At the time

I began working at Four Corners, three machine operators had previously been laborers and one had once worked at the scale house; moreover, one of the mechanics had also begun at the scale house, and Bob had been a laborer and an operator at one time. While only very few employees think they are likely to get promoted to a managerial position, a considerable number seem to expect some kind of promotion. The last laborer fired before I began my job felt entitled to a position as a mechanic, so much so that he openly quarreled with Bob about it, which is said to have led to his termination.

One of the more obvious limits to such internal mobility is gender. Although there have been female laborers and male scale-house operators in the past (with associated incidents of alleged sexual harassment), most occupations are strictly divided according to gender. During my tenure at Four Corners, for example, all three mechanics were male; two (Roy and Jerry) were certified mechanics with years of experience at or approaching middle age, and one (Zack) was in his early twenties. As another example, all three of the scale-house operators—who deal with incoming trucks as they enter the landfill, weighing, inspecting, and admitting them to the site—were women, varying from middle age to college age. And at 4000, the administrative building of Four Corners, located in an old farmhouse on the southern portion of the property, three female administrators handled all clerical work, two of them middle-aged and the third in her early twenties. Workers at 4000 and at the scale house share similar conditions related to ideologies of labor and gender. Both do a great deal of paperwork to record and structure various financial and material transactions on behalf of state regulatory agencies and the landfill company. Even more importantly, both do their work indoors for the most part, removed from direct contact with other, predominantly male, landfill employees. The tendency for women to work indoors at Four Corners is strong enough that there has never been a female operator or manager at the site, or at many others I have visited. When one of the scale-house operators applied for the job of landfill supervisor that eventually went to Doug in 2005, she was not considered, ostensibly because of her lack of field experience outside the scale house, even though Doug knew less about landfills and had never worked for the company.

This class hierarchy is structured around the needs of landfill capital (i.e., the maximal utilization of air space): to bury waste and secure new areas to dump, to maintain equipment and replace personnel as needed, to satisfy and monitor existing contracts and acquire new ones, to prevent regulatory delays and penalties, and to address problems as they arise. As Marx noted, it is the ability of capital to continually provision more workers, more hours, more machines, more contracts (to keep the garbage coming, in other words) that makes it appear as if wage work at Four Corners is derived from landfill capital, rather than the other way around. This dissimulation also affects managerial treatment of the work force, who seem easily expendable and replaceable, depending on what the market in negative value dictates. During my time there, Bob would complain in front of his staff about having to fire people if Four Corners were to lose its controversial sludge contracts. On one occasion, he did so while admonishing Timer for disobedience. Timer was upset that management was storing sticks in the backyard of the house he rented across from the landfill. The wood was meant for fires used to mitigate odor as needed, but Timer had begun burning it himself in protest of a failure to relocate the pile as he had repeatedly requested. Bob warned, while publicly "bitching him out," that six operators had to be fired over the next four months to make up for lost revenue, implying that Timer could go with them. According to some members of management, the loss of the sludge contracts would be partly compensated by the reduced costs of odor control and regulatory fines. Doug even believed that the sludge contracts were costing the company money. Whether or not this was the case, when the contracts were eventually discontinued, Bob was true to his word. The point of his repeated warning, often uttered when distressed or upset, was partly to shift blame away from himself and the company and onto the wider market in waste services.

It is advantageous to managers and owners if employees accept the story that a profitable company ensures the prosperity of its workforce.[4] As I will discuss below, many landfill workers were very familiar with these abstractions, as homeowners financially and morally invested in theories of profit that promised returns of equity and security. Whether acceptance of Bob's market justification was widespread is questionable. Some workers accused Bob of using the sludge as an excuse to threaten or fire people he did not like, in the process assuming more control over the landfill production process. Criticism of Bob's managerial discretion, in terms of hiring and firing, was linked to accusations of favoritism. "Bob's boys" were operators and laborers who were allegedly treated better, had more privileges, and were more likely to socialize with Bob outside of work. To my surprise, some of my coworkers claimed that things had been much better under previous leadership, and it was Bob—not the managerial structure as such—that was the problem. In order to "tell the truth about this place," one mechanic cautioned me, I needed to write about what it was like working for "our Hitler." These criticisms certainly refuse the dissimulating representational tendency that would reduce workers to mere byproducts or collateral damage of market dynamics. On the other hand, by personalizing their critiques of company policy in this way, through their attacks on Bob as an individual, my coworkers failed to relate their experiences to the accumulation processes of landfill capital as a whole. Rather than criticizing landfill capital, workers are encouraged, as we will see, to develop antagonistic relations with one another. Internal labor markets, which allocate wages and benefits by promotion or seniority, only exacerbate such antagonisms.

Class divisions clearly structure relationships at Four Corners, but they do not exhaust the realities of work and life at the landfill. Relationships of stigma, of derogated personhood, reassert the concrete reality of dirty work, which challenges and transvalues the promised transcendence of money. If class struggles and aspirations are about the ability to transcend through value, then such ambitions routinely run aground of the immanent and intimate, at work and at home.

AMBIVALENCE AND CLASS ASPIRATIONS

In chapter 1, I detailed how waste work exposes people to bodily pollution, which they must learn to manage lest they bring traces of their occupation home with them. It is for this reason that landfilling consists of *dirty work,* or labor that is denigrated as undesirable and harmful to one's social identity.[5] As a dimension of human suffering, *stigma* is both singular and commonplace. Most sociologists describe stigma as falling into three types, according to whether a person is stigmatized on the basis of poor character, physical abnormality, or group affiliation. This is what makes work with garbage different from prostitution or law enforcement, for example, which are tainted instead by the moral impropriety and social risks they respectively entail. Labor at places like Four Corners is especially stigmatizing because it can just as easily impugn one's moral character and social status as it can the purity and integrity of one's body.

To understand this claim, it is important to consider not only that relative dirtiness is a matter of interpretation, but that such interpretations are derived partly from the real conditions through which people become the kind of dirty workers they are. Medical surgeons do what most would consider filthy labor and, like landfill workers, must learn to both manage their disgust while at work and purify themselves to transition between work and nonwork contexts. The difference between surgeons and landfill operators is not only the specific type of work performed, but the preparation and investment involved in acquiring these different skills. It takes skill to operate a bulldozer or a compactor, and it takes physical endurance to pick paper all day, but the skill of a surgeon comes with costly and prestigious credentials, including years devoted to higher education and junior residency, as well as successful performances in highly risky, life-or-death situations. Furthermore, any act of surgery, though commonplace from the perspective of the surgeon, is bestowed with singular value through its association with the whole life of a specific patient. By contrast, even the most vital service provided by waste removal and disposal is inextricably collective in meaning because of the progressive anonymization that our waste undergoes as it is managed.

A prevailing individualist ideology in North America represents class membership as an outcome of personal choices rather than historical and social circumstances, and understandings of dirty work develop in line with this interpretative tendency. In order to cope with their stigma, landfill workers both embrace and challenge this ideological frame. In the case of surgeons, their exceptional labor reflects back positively on their person. This is not only about the value of what they do—it is about the widely shared perception that they fashioned themselves into persons capable of performing wonders. Insofar as they are paid high salaries in compensation, furthermore, both the concrete labor of surgeons and the broader conditions that make it possible are represented through the limited and abstract filter of exchange value, which supports an individuating and decontextualizing interpretation of their social becoming.

If one *moves up* to become a surgeon, the general assumption is that one *ends up* working at a landfill. Those who work with waste are seen to be infected with its sticky, malodorous, and dirty qualities. If a surgeon gains status by actively cultivating a rare skill, then a waste worker gains a troubled identity by haplessly mixing with substances the likes of which, it is believed, no one would ever choose to encounter. When indicators of class status, like profession, are interpreted as a reflection of individual potential, a causal relation may be assumed to exist between the kind of job a person has and what they are capable of. It is as if through their labor they become waste themselves: worthless and without potential, human shit.

Class Ambivalence

The potential stigma of dirty work did not seem to bother everyone at Four Corners. Consider Corey, the landfill's environmental engineer. His routine work attire—khakis, clean shoes, no gloves, and a short-sleeve polo shirt—demonstrated his lack of exposure to physical labor or contaminating proximity to filth. This is because the primary task of people in his role is the highly specialized, symbolic labor for which he earned his college degree: poring through regulations and dialoguing with regulatory agencies on behalf of the landfill. Corey was known to be slightly squeamish when it came to "germs" and preferred office work to being on site. While other employees would mock him for this on occasion, behind his back, they would readily accept that his profession was "cleaner" than theirs in multiple ways. As with surgery, landfill work loses its stigma where educational assets like advanced degrees serve as indicators of earning potential. This is a distinction that all landfill workers are aware of and that many hope to achieve for their children. Bob often seemed to defer to Corey's expertise on technical matters, even when he clearly disagreed. Some enjoyed teasing Corey and other college-educated people (myself included) for their lack of practical know-how and real-world experience.

Stigmatized groups often develop countervailing interpretations of their activities in order to resist denigration by others. But this can happen only on condition that one first take into account the negative perceptions of others. If someone like Corey is seen as unequivocally middle class, many landfill workers feel betwixt and between class positions. Whatever they earn, waste workers may experience embarrassment and shame if it is believed that pursuing their occupation squanders their true potential. This was the case for Bart, a lead operator and one of "Bob's boys": "It seems like whenever I say where I work it's hard to say, I don't know why. But especially hard when I'll see someone that I either went to school with or a teacher. Because my kids go to the same school I went to, so I'll see teachers that used to teach when I went there and they'll ask me 'so whatcha doin' now?'" It was no accident that the most feared interlocutors were those from Bart's youth, those who knew him before his path had been set and who therefore made him feel the most ambivalent:

> It doesn't seem like I'm very proud of working at a landfill, but I work hard to do what I can for my family. . . . I don't think [other people] realize how much work there is to it, and how big of equipment, and how technical it is

now and . . . we just aren't a bunch of big fat bones sittin' on a piece of equipment waitin' for a truck to dump and let it sit there. Still, it isn't considered to me as a glorified job or nothing, you know, like a lawyer or a doctor, it's just a landfill guy.

In saying this, Bart shifted back and forth between embarrassment and self-defense. There was a sense in which he was perfectly comfortable with who he was, and another in which it gave him a sense of personal failure. He feared that others would not give him due credit, that they would not realize how intricate his job was or how responsible he had been as a provider for his family.

Other operators claimed to have had success in challenging misconceptions other people had about waste work. Rich, for example, confronted his new landlord over the man's apparent prejudice: "He asked where I work and I told him . . . and he kinda [made a face] like he was stereotyping me as a dirty person. I explained to him, I come home clean." But he remained conflicted: "I don't go around advertising it. Most times somebody asks what I do I say heavy equipment operator." Whether such anxiety is self imposed or learned from negative encounters with others, the temptation is to disavow one's occupation, to rise above the dirt. Though not all will admit to having experienced it, most of those with middle-class aspirations have ways of compensating for this perceived injury to status. Below, I outline two ways in which this is accomplished: through financial spending and saving and through investment in one's children. Both strategies involve using money and the things it can buy as indices of social power. This includes investing in value forms that suggest the attainability of middle-class aspirations, including a house, health benefits, vehicles, regular vacations, and, above all, college-bound children.

Buying and Saving Up

I have a champagne appetite but a beer billfold. I like the nicer things.

GEORGE, machine operator

Near the end of the twentieth century, the then new American middle class was being described as caught between the opposite extremes of wanton acquisitiveness and rational accumulation, or buying up and saving up.[6] Landfill operators are aware of the temptations and opportunities of their incomes, and some have attempted to transform their financial rewards into a secure middle-class future for themselves and their families. Others are less comfortable with their finances—happy with what they are able to afford, but unprepared for larger expenses like college for their children or retirement for themselves.

For many operators, taking a job at a place like Four Corners is already a financially responsible decision, not for the size of the wage they earn but for its regularity and perks. In southeastern Michigan, the most common alternatives for people with their skill assets and class background are jobs as equipment operators in the construction business or factory workers in the automotive industry. The latter is not an option for many, as it is well known that "Fords" and other companies have been attempting to phase out their entrenched unions for decades. Consequently, anyone hired for a high-paying, secure job is typically related to a manager or a high-ranking union member. During my research, wages for construction work were comparable to the base pay at local auto plants, as much as $25 an hour or more, compared with $18.25 to $19.25 an hour at the landfill. However, construction is irregular and seasonal. So, while construction work offered between $5,000 and $8,000 more a year than the median family income for the area, the operators' pay, combined with their spouse's wages, gave them enough to pay for a mortgage on a house (around $100,000 in value, on average) and a vehicle or two. Most importantly, from their perspectives, work at Four Corners offered a good benefits package that many needed for their children and spouses.

Mechanics also could afford houses on their union salaries, and Zack bought one that his fiancée later moved into. Laborers, on the other hand, started at just over $9 an hour. This was considerably higher than minimum wage in Michigan, which was $7.40 an hour, and over time they earned incremental raises so that senior laborers, like the brothers Mac and Timer, made as much as $12 or slightly more. If saved properly, this is enough for a new vehicle—such as Mac and Todd drove—but it is still not enough to acquire many of the standard commodities associated with a middle-class lifestyle, especially a house. Purchasing a vehicle was not only a large expense, it reflected one's identity. Mac and Timer claimed that Eddy's new car had "changed him," that now he was acting superior to other laborers. It was around this time that he also began working more with Todd, doing higher-prestige jobs like tending to the gas wells and odor control system. Some laborers supplemented their income with other work, usually as part of the area's informal economy. It might not be realistic for them to aspire to owning their own house, go on lavish vacations, or send children through college, but they did endeavor to own their own vehicles, live independently, and support their families as best they could.

Being the sole income earner or provider for one's family involves heightened financial awareness and anxiety. Each worker, in a sense, becomes a capitalist investor in their own wage labor, which is a means to generate more wealth and class advancement, rather than a source of collective identity. This was an explicit goal of the neoliberal social movement, also known as "freshwater economics," that developed at the University of Chicago in the midtwentieth century. Before their ideas inspired the Washington Consensus and the North American Free Trade Agreement, influential neoliberal theorists like Milton Friedman and Friedrich Hayek targeted the perceived economic inefficiencies and political dangers of the powerful unions left over from the Fordist era. Destabilizing the power of unions was not only about enriching capital, but was tied to Hayek's belief that individuals had to be taught to be economically rational.[7]

Despite, or perhaps because of, a widespread tendency to associate the history of Michigan's meteoric rise and fall with Fordist-style industrialization, many Michiganders are themselves critics of the excesses of this period and the general legacy of "Fords," as it is popularly known. In fact, the increasing rarity of high-paying union jobs with good benefits has fostered anxiety and resentment. In my experience, Michiganders often blame the decline of the big three auto companies on the greed of corporations and auto unions, the betrayal of consumers who no longer buy American, and often all three. In southeastern Michigan, it is therefore not sufficient to label the transition to neoliberal post-Fordism as merely the imposition of a new regime of accumulation or governmentality by societal elites. Rather, this is a far more ambiguous and homegrown product of ongoing struggle over the meaning of work and gender amid the ruin of a once prosperous era.

Though operators and mechanics at Four Corners were members of a union, none that I knew of took union membership as seriously as they did the conduct of their own financial affairs. The primary way in which they expressed their financial agency was with regard to overtime work. Though effectively limited to fifty hours a week, managers occasionally let those desperate for overtime work extra, particularly when there was a shortage of workers or a push to finish a special project. This gave operators a chance to occasionally increase their weekly wages and a sense, however limited, of financial control. The extent to which middle-classness is integrated with systems of financial credit and reflexivity is evident in Bart's description of his monetary habits:

> I'm not that great of a money saver. A good money spender, and I'm a good bill payer. . . . I've got awesome credit. I could go to a bank and probably buy about anything there is 'cause I don't ever remember being late on a bill, ever. So that's good, but I'm great for a bank because I finance everything there is! It don't matter if I make twenty thousand or if I make a hundred and fifty thousand. . . . I know that's not good but that's the way I am.

Bart associated his responsible bill paying with having good credit, made possible by his regular income from the landfill. Yet his habit of financing large purchases, such as family vacations, simultaneously made his earned income, whether "twenty thousand . . . or a hundred and fifty thousand," somewhat irrelevant. The more he paid off his debts, the more his credit allowed him the opportunity to spend however he liked, leaving him at risk of overextending his family's finances. The push-and-pull that Bart experienced, between buying up and saving up, was not simply related to his family's individual spending habits—credit had been actively promoted by North American lending institutions, and U.S. consumer debt rose from 110 percent of personal annual disposable income in 2002 to 130 percent in 2005, shortly before the recession.[8]

There is less at stake, and presumably less need for monitoring and extending one's finances, if an aspiring middle-class worker does not (yet) have the responsibilities of a family or has avoided paying child support. At Four Corners, the youngest employees are typically the least permanent, holding part-time positions at the scale house or as laborers. This may be because they have not yet acquired the skills to work as mechanics, operators, or administrative staff. On the other hand, they might desire a more temporary or laid-back job, one that requires only minimal skill and provides a weekly paycheck slightly over minimum wage. But younger people are not always in less permanent positions at the landfill. Zack began at the scale house as an adolescent and later moved into the shop as a mechanic. For him, any stigma associated with waste work was vastly outweighed by the possibilities it provided. The house Zack was able to purchase was something nearly impossible for friends his age. The stigma of waste work may not be experienced as such until he is old enough for it to be recognized as *his profession.* When I knew him, Zack could still fantasize about applying at the local prison as a guard or finding an altogether different job.

Other landfill employees are not so fortunate. Being older, they have become enchained to social networks that invest them more deeply in what they do for a living. Their possible futures are more closed, overdetermined by their present obligations. A long-term operator like George, for instance, had gradually become resigned to the fact that landfill work was his career. He didn't plan on doing anything else until he retired, nor would he necessarily want to. As with many of his colleagues, accepting waste work as his profession was not exactly a choice for George, but a consequence of accumulating financial and familial responsibilities. For those I interviewed, becoming a spouse or a parent meant having to be responsible. This entailed working hard to bring money home and managing it to accomplish the greatest good. Though Zack was not similarly enchained by his own relationships, he also received none of the respect and personal satisfaction that come with being seen as a successful provider.

For landfill workers earning middle-class incomes, such as operators, becoming responsible forces them to give up lucrative jobs for more steady, long-term work. Many feel obligated to settle into their current profession because they believe they are choosing stability and security over what George described as a more volatile path of living paycheck-to-paycheck. However, stability is not only about the balance and regularity of a steady wage; it presumes a moral character with respect to finances and relationships. Responsibility means sacrifice—frivolous purchases must be abandoned, or perhaps set aside until retirement (in chapter 3, I discuss how some desired items could be procured by other means, precisely so that they did not have to be bought with one's earned wage). The preference is not for saving over spending per se, but for saving up to buy things that are seen to be, in and of themselves, secure and responsible. Houses in good neighborhoods and brand-new vehicles are two expenditures that most operators are able to afford because of their steady incomes. These are responsible purchases because they are intended to save and even to make money over time (by accumulating equity, for example). On another level, they are believed to create a material and symbolic foundation for the growth of a family and its future.

Some landfill workers could afford such things before they began working at Four Corners, but others owed their responsible buying to their choice of a regular paycheck. According to Carl, a younger operator who began as a laborer, the "big pay" he now receives brought him a middle-class lifestyle he never would have imagined otherwise. For the first time, he and his wife had a house, a driveway, a basement, a pool, and a two-car garage. It was the same for Bart, who contrasted his current circumstances with those of his childhood: "My parents were kinda like I was, they had vehicles that got us by, but they never had brand new vehicles." Unlike his parents, Bart was able to buy a house at a young age and take regular vacations with his daughters to the ocean: "I never remember doin' much with my parents when I was really small." Vacations may not be an investment, but they signify a middle-class lifestyle of earned leisure. And they, too, are implicated in the renewal and preservation of family relations, through shared experience and memory.

Other operators might label Bart's decisions frivolous, as he was well aware. Not because he was spending money, but because he didn't know how to save up for what he *should be* spending money on. The slightly older George is a good example of one who focused almost exclusively on saving up for a secure future. He also contrasted his own relationship with money to that of his parents: "Like I say, my parents were not very fortunate and my dad was not very business smart. And so I said I don't wanna end up like that." After several years in the waste industry, George had saved up enough money to purchase a new home shortly after he was married. He and his wife agreed that they ought to "build up a bank account" before having children, so they waited seven years to do so. George was clearly proud of the results: "We weren't knocking down big bucks, but . . . there was enough there that there weren't great favors and that's how we did it. No help from anyone, no help at all."

What enabled George and his wife to accomplish their goals was only in part a consequence of the financial decisions they made. He said that he learned a great deal from his father, who worked in the same electric company most of his life, about the importance of settling into a job. George called himself a "lifer," not because he loved working at Four Corners (he would always rather be home), but because he believed in the opportunities that come with holding a steady job. It was being a lifer that allowed George's wife to quit her job and raise their two kids at home, and that kept him motivated with dreams of an extravagant retirement. George had long planned to retire at fifty-five but was now aware he would not be able to, because of the impact of the global recession on the stock market, which he had been "playing" for over twenty years. This did not lessen his excitement about the prospect of no more work:

G: I've been planning this for a long time, I don't know how it's gonna work out. Let's put it this way, I got big plans! I just wanna be financially secure; I just wanna have it made at the end. All this hard work I want it to pay off.

J. R.: What qualifies as financially secure, in your opinion?

GEORGE: Have a house of your choice, maybe a race car for a hobby, and just being able to do things, you know be able to go out there all the time, go on nice vacations. Just finally being able to do what I want to do, instead of just common work everyday.

Whether security means being able to give vacations to one's children or saving up for a dream retirement depends in part on the biography of the worker and whether their children are grown, awaiting college, or yet to be born. But most seemed to believe that there was an ideal balance between saving and spending that they had not quite attained, an ideal that seemed even more unattainable after the financial crisis of 2007–8 (also arguably an outgrowth of neoliberalism).[9]

Anything They Wanna Be

Buying is unable to secure middle-class status without saving. Similarly, giving your family the *things* that you never had is less successful as a strategy than giving them *opportunities* for a better future. For employees at Four Corners, as for most North Americans aspiring to the middle class, encouraging and imagining success for their children offers a sense of personal fulfillment. More specifically, what is hoped for is that one's children will have the freedom to choose to do "whatever they want." The ambiguous position of middle-classness—more aspiration than actuality—hinges on the appearance of free self-creation. Money is nothing but the objectification of such agentive possibility, of the option to do as one pleases. The problem is that not just any life will do, not every kind of job indexes such freedom. The unpleasantness of dirty work suggests that no one would rightly choose to do such a thing if they had the power to do something else. If the children of waste workers acquire positions as college-educated professionals, that is, if they manage to be unambiguously middle class, they will signal to others that they had the accumulated assets to be what they wanted to be. This presupposes that their parents gave them more than things, but empowered them to create themselves anew. In this way, a middle-class identity is ensured through what it makes possible in the next generation. Waste workers imagine a future that *will someday guarantee* that they were always already middle-class.

But there is an obvious paradox at work here. If a prevailing individualist ideology holds people accountable for their own success or failure, judging them accordingly, then how can the success of a child reflect back on a parent? In a sense, this is no more contradictory than the historical coexistence of individualism and Christianity in the United States, which similarly combine individual responsibility with ultimate dependency on the grace of God. Specific acts of interpretation can render individuals as isolable moral actors responsible for themselves or, by contrast, as partible persons defined instead by their relationships and obligations to others.[10] For example, the ideological separation of home from market, to which I will turn in chapter 3, facilitates seemingly contradictory representations, such that a wage earned and bestowed in one sphere can make dependency manifest in another. A dirty worker may earn a tainted wage on behalf of their family, but only they are polluted as a result. A newly middle-class professional may pursue higher education to earn a pure wage, but by transcending their social inheritances they also demonstrate their indebtedness. If a child succeeds, she may do so both in spite of and because of her parent.

Given this contradictory and complementary double bind, hope for the future may easily mix with regret for the past. Dan, for example, was not pushed to go to college, and among his siblings, only his youngest brother ended up attending. He imagined a better life for his own son, which he viewed as a step above the many blue-collar professions he had worked in:

> I wanted him to go to school so he wouldn't have to work construction, be a truck driver. . . . I've always made good money and I've been a good provider, but get educated and do what *you're* doing [referring to the anthropologist], do something with your mind and stay clean . . . make your generation a little bit better, live longer. . . . I'm about ninety-nine percent sure that it's taken a few years off my life, one way or another.

Dan envisioned an unambiguously middle-class life, involving clean, symbolic labor. This was not only about money per se, but about building a better, longer, healthier life. It was about always having the option to do otherwise. Dan used his gendered success as a provider to justify the legitimacy of his profession: "I'm an operating engineer that pushes garbage, okay, making damn good money. . . . I'm here to support my family same as [other people] are." His aspirational middle-classness was evident in his effort to be responsible, to work hard to support those who depended on him.

Unlike Dan, Bart had attended some community college. In fact, a high school guidance counselor encouraged him to pursue math and accounting, which he majored in before switching to auto mechanics instead. While it was clear that Bart didn't regret changing degrees, he did like to know that he was capable of leading a different life, perhaps a white-collar one, given his natural proficiency with numbers. Moreover, as with all the parents I spoke with, it was important to him that his daughters choose work that appeals to them, whether blue- or white-collar. But the possibility of doing whatever you want implies having the power to choose, and that power is popularly imagined to come from earning a college diploma:

J. R.: Do you expect your kids to go through college?

BART: I want 'em to. And, the oldest one, she's planning on it and I expect her to.

Bart described how contributing to his eldest daughter's postsecondary education would have been unthinkable without his current income:

Where I worked before, I, of course, would try to do anything I could to help out to put her through college, but I don't see how I could do hardly anything at all. But, making more money like this I would find a way to finance something to where I would get my kids through school. [In the past] they'd probably have to pay for more or less everything by themselves.

The more college is paid for by parents, the more children begin life unencumbered by financial debt. Such freedom from constraint is the ultimate goal. College is another financial burden associated with being a good provider. In a way, it is the most successful investment because it has the potential to renew a family's class standing (and therefore the provider's sense of accomplishment) into the indefinite future.

That children should be unencumbered by financial limitations does not mean that landfill workers do not have their own preferences as to what their children *should* do. While all parents that I spoke with said they would be happy for their children to work in the waste industry *if that is their choice,* very few would prefer they make that decision. According to Bart:

[My oldest daughter] can do bigger and better things. I'd try to tell her workin' outdoors, the mud, the rain, the snow, the cold, you do get dirty, you do get stinky. I would think that she could get herself a nicer job. The younger one could too. Dream world would be my daughter bein' in some type of medical field, and the youngest one I always thought she'd be good at like a vet.

Bart's preference was not for just any professional, middle-class occupation, moreover, but for a secure position: "The older one, she actually thinks she wants to be a lawyer but I think that's a tough field; there are tons of them out there. If she would specialize in something in the medical field you can do pretty good." Both veterinary and medical practice can be dirty and involve unpleasant work conditions. The goal is not to be completely free of filth, but to achieve something that suggests achievement and self-creation.

Bob wanted the same thing for his adopted son, Taylor. So he was ambivalent about the child's professed desire to do his father's job one day, which he showed from an early age by creating multiple drawings of pieces of landfill equipment. Bob treasured the pictures, which were hung above his desk in the management offices, but he related his son's desire to be a waste worker to a lack of interest in classroom pursuits (something his father shared at his age): "[He] talks about working out here someday, running equipment. . . . He has a lot of problems in school. . . . I hope he goes to college someday, but . . . he has a hard time goin' to school, sittin' still all day." This inability to sit still was something Bob could relate to, and in a sense he was proud that Taylor would rather be active and outside. But at the same time, this was not Bob's professed desire for his son's future. Rich had similar thoughts with regard to the future of his children. According to him, they would do well in the waste industry, but he feared that it was not stable enough to support them: "I don't think it's the most secure of trades. I'd try to steer them away from it, get, um, something that's a sure bet." This suggests that the ideal is, in fact, to create more responsible, middle-class providers. Parents do not want their children to have it easy; rather, they hope for a life that will build character, one that involves responsibility, integrity, and a work ethic most of all. Bob, himself a former operator, expressed concern that his child not grow up "spoiled" or with a false sense of security: "If he wants to be an operator that's fine, but he's gotta earn it, I'm not gonna give it to him. . . . I don't want him to have that false sense that he's got something guaranteed here." It is important for children to think that anything is possible *and* that nothing is guaranteed.

Self-creation can be signaled only if children choose the *right* profession, one they can be seen to have earned and to treat with care—thus proving that they became the people they wanted to be. The ideal, in other words, is for children to have had a choice and to have made the most of that choice. Even the intuitive middle-class American remark "I want my kids to do whatever will make them happy," which was regularly repeated by my coworkers, implies *the power to choose between unhappiness and happiness*. In order to build a middle-class legacy, however, landfill workers hope that choice will lead to something better than their own lives. As Timer told me, if his son expressed interest in working at a landfill, he would be completely frank: "Landfill ain't no place to work, I don't want you to wear the same shoes I wore. You can do better than that." Wearing different shoes (quite literally, loafers rather than work boots) does more than help the next generation live better, it helps their parents get beyond what can be a strongly felt sense of personal limitation.

Perhaps for these reasons, young people at the landfill, like Eddy, Zack, and Todd, were sometimes spoken about as the worst examples of what children could become. More senior workers complained when they did not show sufficient respect. Todd once said to me, unsolicited, "I wouldn't want to be anyone at the landfill," and that he planned to continue going to college precisely to avoid their fate, adding that he would just sit in the back of the class if he smelled like garbage. People in management encouraged all three to continue their education but also looked to promote them and grant them more responsibility to reward signs of ambition.

In my experience, Eddy, Zack, and Todd would all regularly criticize management and other workers, particularly for perceived slights and inconsistent treatment. It is commonly said in North America that children and young people do not seem to appreciate what they have, meaning that they refuse to acknowledge the role of others in their self-creation. The problem with hoping your family doesn't have to experience life the way you did is that over time, it may become difficult to relate to them. Getting an education has long been associated with middle-class aspiration, but it also creates a paradox for the working class, potentially imperiling their sense of dignity and self-respect.[11] For farmers and autoworkers, it may be preferred that children enter the same line of work as their parents. Many of my coworkers at Four Corners aspired to be autoworkers or farmers at one time or another, and they were fond of pointing out that money and education might actively preempt independence, common sense, and good character. On one occasion, Roy confronted George about the latter's son in the break room. Roy had tried to get his help with his home computer but received no response.

George responded by lamenting, "You spend all this money thinkin' they're supposed to learn!" "They don't!" another operator quipped. Some complaints and concerns about children suggested implicit approval, as when a child would monopolize the home computer "for school," but of most concern were those that impugned character.

For a time, I was a popular target for such concerns. As a person who chose to work at a landfill as part of my graduate education, my presence was found at times amusing, perplexing, and suspicious. Landfill management, particularly Corey, were open to the idea of inviting college students and professors to their site, whom they believed would be more open-minded than reporters, environmentalists, and NIMBY locals. I stayed longer than most, however, and was intent on "being treated just like everybody else," as I often put it. My very first week picking paper along the road, I was greeted with shouts and laughter by passing coworkers: "So you wanted to work at a landfill!" "If I was a anthropologist, I'd work at a strip club . . . in Hawaii!"

Whenever I was unable to perform a task, due to either ignorance or physical weakness, my superiors would chide: "They don't teach you that in grad school?" I received other, similar insults—my favorite being "He throws like an anthropologist!"—most of which signaled playful acceptance of my presence on site. On other occasions, this led to discomfort. There were those, whom I do not discuss in this book, who refused to participate in my research. Even people who had agreed, like Bob, would sometimes reveal anxiety about my being there. In response to someone who asked why I was there, Bob once replied, "He's here to find out why us retards are so retarded."

My ambivalent reception at Four Corners says as much about the distance between researchers and the researched as it does about the equally ambivalent relationship many of my coworkers had to people from outside the landfill's orbit, including their own children. It is not a stretch to suggest that some related to me in a parental way, trying to give me guidance or seeking advice about their own children. On one occasion, one of the managers asked me about how to get grants for undergraduate education. His daughter was hoping to transfer from Eastern Michigan University to an expensive art school in Chicago where books cost $1,200 a semester. He told me that she planned to go into marketing and was clearly relieved when I said that people could make a good living doing that. Precisely because they had so much economically, symbolically, and emotionally invested in their children's futures, my coworkers worried about their children being ungrateful, undisciplined, and inexperienced.[12] At the same time, they also worried about their own jobs and their ability to perform them adequately.

Second Skins

George was proud of his middle-class lifestyle, but aware of the lingering stigma of his occupation. During an interview, he expressed his ambivalence to me: "People probably see my house and don't realize who lives there. That's why I like to have nice things, that's why my wife and I like to live next to upper-class people: just 'cause I work at a dump doesn't mean I'm a dump!" George contrasted the wealth and success of his home life with his occupation, claiming that what he did for a living did not diminish who he could be or what he was capable of. If enclosures are additional skins that envelope the body, then we can imagine the continually changing exteriors and interiors of homes as both a permeable container and a medium for display, sheltering bodies while also telling others what kind of people they are. Not only could George and others escape the literal contamination of the landfill by going home, the home itself projected a clean image.

Waste disposal cleans all of our images. Most dwellings are expected (at minimum) to provide occupants with the capacity to smuggle out garbage in bags and cans and to pipe the most unsightly waste underground. Boundaries between work and home in North America have always been more ideal than actual, particularly for lower classes and ethno-racial minorities. The gendered North American divide between *housework* and *working on the house* is an ideological separation supported by the post–World War II reinforcement of patriarchal and racialized privileges and a corresponding transformation of the ideal home. Yet the related distinction between residing, or being at rest, at home and laboring at the workplace remains tied to middle-class aspirations, even as an impossible ideal. Like many similar suburban neighborhoods, the recently built, upscale housing development that I call "Silent Pines" was considered a "bedroom community" by local residents, a term that signifies nonproductive and leisurely occupancy. Homes are not places of rest, but places whose reconstructed sameness—that is, transcendent separation from the base realities of time and process—must be tirelessly maintained through cleaning and repairs. Landfill workers are thus provisioned with a refuge from their labor, one that is rented, mortgaged, or owned by means of the abstract labor power for which they are paid; that is made possible by the collective subtraction of transient materials by means of the concrete labor people like them perform on behalf of all homes; and that is actively maintained by means of the informal work that members of their household routinely do in and on the home itself.

One younger operator, Dan, lived in a distant, suburban subdivision down the highway from the landfill, with his wife and young children. He "bitched" about his neighbors, their lack of sociability and their peculiar complaints about his dog barking in the middle of the afternoon or the location of his mailbox: "I'm not built for subdivisions," he told me. Dan hoped to move farther into the country one day soon, once he had improved the upstairs. Like many North American homeowners before the housing bubble burst, he was confident that it would sell at a profit once he made necessary improvements to the product itself. Like George, furthermore, Dan's home showed no sign of his profession and he told me that, while he did not hide what he did from them, few of his neighbors suspected what it was. On one occasion, I was invited with others to Dan's to play poker and watch the University of Michigan play football against their bitter rival, Ohio State. Upon entering, I was politely asked to remove my shoes, and it was clear that Dan's wife, Barbara, insisted on keeping the house very clean, at least when entertaining guests. At one point during the game, one of the children tracked dirt onto the living-room rug from outside and Barbara became instantly irate. Dan promised to clean the mess but did not do it right away. Moments later, when their youngest began crawling toward the dirt, Barbara yelled at Dan from the other room to help: "The baby's eating mud!"

The middle-class call to subtract transient materials from the home is about maintaining freedom from unwanted possibilities. But such anxious freedom is neither possible nor desirable for all workers at Four Corners. During its history, a small number of operators, managers, and laborers have lived much closer to the landfill than George or Dan or I did. Dan's work boots were not visible in his home, nor was his uniform, and, unlike some of his coworkers, he displayed no trinkets recovered from Four Corners in his home. One can live in a subdivision and maintain a home to project a desired image. Even so, transience can never be left behind entirely. Not far from Dan's street is the coast of Lake Huron, where signs warn against swimming or fishing because of contamination. And on the horizon, one can see the outline of a massive nuclear power facility (the distant presence of which worries some local residents almost as much as Four Corners).

More generally, Dan's house, a perceived asset, likely declined in value along with many others after the housing bubble burst under the weight of toxic subprime mortgages. But this is not the only investment associated with middle-class aspiration.

ACCIDENTS AND ESSENCES

A wage represents a company's abstract interpretation of a person's capacity for work, but this capacity is differently interpreted in concrete work interactions. Being perceived as a bad worker, like being stigmatized as a dirty one, involves interpretative assumptions about what kind of person you are and what you are capable of. While getting dirty could not really carry stigma within the orbit of Four Corners, being perceived as incompetent or accident-prone did. Employees who could not do their job well, routinely injured themselves, or damaged property were in danger of losing their job, but before that they would lose the respect of their peers.

In fact, some of those who were stigmatized as incompetent were known to act squeamish about getting filthy. One who can't smell the money in the waste may also have trouble tipping and compacting it. A group of the younger operators would mock George behind his back for being sensitive to smell; at the same time, he was criticized for being older and having "lost his touch" with the machines. Usually those engaging in the mockery were not only younger operators, but also former construction workers who prided themselves in having superior skills in operating machines. One instance I documented occurred while operators were transitioning on and off machines between shifts. While in the break room, the mechanics explained that they had been told something was wrong with one of the dozers, known as the "Nine." One of the operators shifted the discussion from the possibly faulty machine to George's allegedly faulty labor. "He can't see right on the Nine, but the Eight was much worse," one said. "Boy it's hard to see on them Nines, ain't it?" another mocked.

It was no accident that George was singled out for criticism, given his relative seniority in years. As far as the landfill company was concerned, the time he spent on a dozer was worth more than the time of an operator his junior. Again, wages are paid not for concrete labor, but for abstract labor time—that is, for the generic capacity for work. If one of my coworkers worked particularly hard one day, picked more paper, or compacted more waste loads, for example, they would still earn the same amount as someone in an equivalent position. People were occasionally punished for working badly and earned less, however. After Todd and Eddy accidentally burst a line for the methane gas, Bob was furious and sent them both home, thereby imposing a somewhat arbitrary financial penalty as punishment for the mistake.

In many occupations, the only ways to earn more are to make a higher wage by moving up in the informal labor market or to work more hours and possibly get overtime pay. In real time, some people work faster or have better vision, but a wage is calculated on the basis of a linear and homogeneous conception of temporality. The complexity and depth of work relationships are reduced to what they earn at the end of every pay period. At Four Corners, wages were calculated on the basis of a punch clock. As at other sites of industrial labor, the laborers, mechanics, and operators had to "clock in" at the start of their shifts and "clock out" at the end. With the rise of clock time in industrial capitalism, the exchange of a wage for a precise quotient of expended labor became associated with the work ethic and a measure of character. The punch clock at Four Corners was not merely a device for wage calculation; it made individuals responsible for their own time-discipline. We were warned never to clock in anyone else using their time card, though people would do so on occasion. In this way, objectified notions of time and abstract labor become associated with moral reflections on personhood and masculine identity. My coworkers would both submit to and make creative use of abstract measures of their labor to frame their everyday workplace interactions.[13]

Safety as Work Discipline

Human error could be regarded as inevitable—as it is often depicted in science and engineering, for example. But with the increasing prevalence of neoliberal forms of governmentality in work settings, injury has been reimagined as something that can be avoided through the cultivation of worker responsibility. Since the nineties, employers in North America and throughout the world have focused on training their workers to take it upon themselves to act more safely.[14] Before joining America Waste, would-be employees must watch safety videos about work hazards and complete a short multiple-choice exam. The one question I got wrong asked who was responsible for my safety while on site. I chose "the safety manager," assuming that there was such a position. On my first day, Bob was required to discuss the question I had missed. He explained that there was no safety manager and that I alone was responsible for my being safe on site. This was indicated in other ways too. A sign in the maintenance building, hung prominently between the management offices and the break room, displayed how many days had gone by since an employee injury. It was suggested to me by management that this was meant to encourage workers, so they could take pride in what they had accomplished. However, this sign could also be seen as an additional measure of abstract labor time, one uncontaminated by injuries to actual laboring bodies. Every so often, America Waste paraphernalia was distributed as a reward for our good safety record. One time, Bob handed out hats to everyone, one at a time, "here, here, here" with a flat monotone. There were not enough to go around, and he joked with me, "No, I don't think you're safe enough," but seemed to feel bad about not acknowledging my officially accident-free record.

By implication, accidents are the result of individual worker negligence rather than circumstance or company negligence. It is on this basis that Four Corners garnishes their workers' wages for the vacation/holiday fund. This financial regimen is meant to compensate for not having paid vacation or time off from injury. Money for the fund is automatically extracted from weekly gross earnings at the rate of 14.5 percent of one's paycheck. This makes operator earnings tight during the middle of the year, but at year's end, after taxes, it can amount to as much as $7,000 to $8,000 in reimbursement. In this way, the landfill itself institutes a form of financial management under the rubric of protection. The fund presumes worker responsibility for sickness or injury—equating it, along with holiday absences, as time off. In general, time off was something that my coworkers and I tended to represent as sacred because it separated us from work. In fact, as a motivating goal to continually work toward, the promise of time off gives rhythm and texture to productive temporalities, whether falling at lunch, at the end of the day, at the start of the weekend, or just before an extended holiday or vacation. The shared desire for time off provides economic and moral justification for the financial contribution they are required to make. Presumed control over one's absences is a narrowly confined choice, but a real one nonetheless. This reflects a tendency to shift responsibility to the workers themselves as individuals, rather than as part of larger social collectives.

Work and Character

Though the wages we earned were not much discussed, many people at Four Corners readily discussed how many hours they had worked on a given day or week. This implicitly suggested greater financial remuneration for our efforts, but it also was a sign of being capable of hard work, and not just work in the abstract. It has long been noted that Americans tend to equate personal character with hard work. Timer was very clear that he did not want his children to do what he did for a living, but he once paid his eldest son to come pick paper with him, bragging about his work ethic. This mattered to Timer, as it did to many of his coworkers.

The opposite of a hard worker was a lazy or incompetent one. While interpretations of what this meant varied, the figure of the hapless, accident-prone fool was its clearest expression. There were stories about workers who sought to get injured in order to earn compensation, or "worker's comp." One former operator, Arnold, was criticized for allegedly doing this. When he and Bob were both laborers, I was told, they were close. Some said that Arnold would even clock Bob in and out so that he could "goof around" instead of picking paper. Bob was promoted, while Arnold became fixated on getting injured. When I asked what happened, one employee joked that Arnold cut a branch off a tree he was sitting on, which they likened to a cartoon, and then "broke his back." In conversations with me about safety, Bob alluded to a former operator—who may or may not have been Arnold—who injured his hand to get worker's comp. According to Bob, he had tried to inspire this man to avoid accidents by referencing the number of days the landfill had gone without injury, prominently advertised in the maintenance building. "That's not for us," Bob told me, "it's for them guys!" However, none of the workers with whom I talked took much pride, or had much interest, in the overall safety record, which they typically associated with managerial interests.

Working at a landfill can be very hazardous, particularly for people who are walking around on foot with big machines all around them. This awareness of danger may have been sublimated into other concerns, as I recorded in my field notes:

> It occurs to me that Eddy and I talk about nuclear war, global warming, another ice age, all while facing risky work situations we don't reflect on. [During that conversation] I was concentrating on not getting hit by a truck, not wandering too far from the blinking light on the flat bed, looking busy—which means not picking too close to Eddy—and looking for steel or big pieces of shit, all the while trying to keep my footing on the spongy ground.

Our managers were aware of the risks. On foggy days, I was warned to be particularly careful; on cold days, I was reminded to wear appropriate attire;

on hot days, people might bring us beverages. But it was also clear that we were ultimately responsible for our own personal safety. Knowing this, workers may obfuscate their responsibility for accidents through the accounts they circulate, blaming machine error, other workers, or inauspicious circumstances.

After about a month of working at the landfill, I was involved in my first near-accident and didn't even realize it. After I had been picking the exit ramp for a few hours, a bulldozer pulled up next to me. One of the operators, Aaron, leaned out and said, "Want some safety advice?" He told me to be careful of picking the south ramp in wet conditions because it can be hard for drivers to control their vehicles as they descend. "You can be driving straight, but your trailer might swing on its own." If this happened, the driver could accidentally hit me in the process. Moments before, Aaron had started going straight down the ramp when he temporarily lost control and his vehicle began going sideways. He had shouted something to me as it happened, before regaining control, but I hadn't heard it. As Aaron recounted this near-accident, I felt slightly defensive, both because I did not like thinking that I might have been hurt or killed and because I resented the implication that if this had happened, I would have been responsible. Conditions on the exit ramp were such that, even if I did notice in time, I might not be able to move fast enough to avoid injury if an oncoming truck lost control. But I tried to act grateful for his help. He may have taken notice of this, because Aaron shifted blame further away from himself before concluding our conversation: "Some of these Ay-rabic truck drivers can't hardly go in a straight line, let alone maintain control of their trucks." I thanked him, though I didn't admit any wrongdoing. In our parting exchange, the conversational footing became more humorous and playful:

AARON: I just don't want to see a red stain on the road!

J. R.: And I don't want to be one!

And he drove off. A few weeks later, I learned that Aaron had been telling people that I was almost killed while picking the exit ramp because I hadn't been paying attention. At least, this was how some of my fellow laborers relayed his speech back to me, evidently looking for a reaction. Secretly offended by their taunts, I recorded the outrage they had hoped for in my field notes:

I am annoyed that [Aaron] is spreading the rumor that I almost was killed picking at the exit ramp, because I worry it makes me seem incompetent.

[When I hear it] I typically respond by downplaying the danger and emphasizing my awareness of the situation, then back-pedaling and remarking that I appreciated his advice, which is what he called it.

At the time, I didn't understand how and why the anecdote was circulating; I didn't yet know that this was part of the normal process of discursively managing accidents and near-accidents at the site. Aaron didn't know if I was going to tell people that he had lost control of his vehicle (I hadn't) and was preempting that by circulating his own version.

Being accident-prone meant that one could not be trusted to get the job done or even to protect oneself from harm. For example, Todd was a regular target of criticism and mockery for his tendency to damage vehicles. Though a laborer, he was given additional responsibilities—such as tending to the gas wells and perfume sprayers and stink eliminators—so he had access to vehicles more often than those sent to pick paper. On one occasion, he damaged the transmission ("blew the tranny") of the GMC truck while trying to get unstuck from mud. Afterward, when he and I were assigned together, I also got the truck stuck in mud while driving but managed to wriggle us out. When I told one of the managers about the incident later that day, Todd began to laugh at me. "You aren't supposed to tell them when you fuck up!" The people around us then began to tease Todd, in response, for always getting into accidents. "You didn't blow the transmission getting out?" one mechanic asked me, laughing. Todd hated being teased in this way and did what he could to avoid his mistakes being discovered. Once, he noticed that we'd lost the oil cap on the car and stuffed a rag into it, remarking that he hoped Big Daddy didn't notice because then he would see an unreported dent and a busted rearview mirror. On another occasion, he forgot to turn off the valve on the water trailer while filling it up. As a result, the back pressure from the hydrant burst the hose that connected the trailer to the perfume lines. He sped over, red-faced, hurrying to shut it off before it attracted attention.

No one was immune to mistakes, and other workers behaved similarly. I was encouraged by mechanics to cover up a minor accident when I dented a vehicle on a pylon. They reasoned that I should not have been instructed to move the vehicle in the first place (since only unionized mechanics and operators were allowed to operate machines) and hammered out the dent so management wouldn't notice. When accidents did come to light, operators might laugh about them. On one occasion, an operator accidentally tipped a pile of garbage, stacked six feet high, onto the top of Bart's dozer; he jokingly said it looked as if a pile of garbage had begun to move of its own accord, cartoonishly wandering around the site. But he also confessed, amid the laughter, that he worried the cabin would cave in, that someone might get hurt or management might learn of his mistake. Finally, he strongly implied that Bart's lights may have been off. As long as this was mere rumor, the details could be freely manipulated.

The opposite of being hapless and accident-prone was not necessarily being safe, which could be seen as cowardly and a different form of incompetence. If done skillfully, being reckless was not a bad thing in and of itself; it could be an expression of skillful manipulation under difficult circumstances. So it's not surprising that Todd imagined proving people wrong not by being safer, but by being spectacularly unsafe. When he got his new truck, Todd talked about wanting to drive it up the east slope of the landfill with the other operators. When I told him that sounded dangerous, he smiled proudly. Indeed, the operators would occasionally reminisce about drag racing down the south road in their vehicles, when one of them hit a large pothole and nearly crashed. They allegedly raced work vehicles to the top of the extra dirt piles lying around. Todd would describe these events with excited glee. Company policy at Four Corners, as stated during one of the regional meetings, was "Don't set cruise control for sixty-two, drive at fifty-five," meaning that people should operate at a safe level at all times. In truth, it was generally assumed that the rules were for visitors and novices. We did not wear hardhats and goggles and drive ten miles per hour, as all visitors were required to do. This was a sign of our distinction from them, suggesting that we were competent enough to know when *not* to follow the rules. If ever our work practices led to injury, however, this narrative could quickly change.

Alton's Accident

The worst accident that occurred during my time at Four Corners befell Alton, an older operator. I had always liked him because he was friendly and respectful. Many operators and truck drivers littered, so laborers would have to pick up after them. Alton would pull over and hand me waste for my plastic bag, saying "Here's something for your garbage," and then offer me water from his own supply: "Here's something for your stomach."

I first heard that Alton was in an accident from another operator, who'd seen him near the ash cells rolling in water to put out fire on his leg and arm. Somehow the methane gas he was working near had caught a spark and shot a flame into the air, lighting one side of his uniform aflame in the process. Afterward, Alton reportedly drove himself to the emergency room, so everyone assumed it could not have been too bad, though one of the mechanics claimed that he had had to drive himself because no one would do it for him. After I learned of the accident, I spread the rumor to the people I ran into throughout the day. Some laughed, some sympathized or judged others for finding it amusing. Timer used it as an opportunity to criticize the company's safety policies: "He'll probably be back to work tomorrow!"

In fact, rumor spread the next day that Alton was dismissed without pay for between three and five weeks, that he had a second-degree burn on his arm and a third-degree burn on his leg. Rumor also spread that Bob wanted to use this as an opportunity to fire Alton because he had gotten into accidents before, typically rolling the water wagon. The rumor began to mutate into an official narrative when Big Daddy intervened. He decided that the accident was not Alton's fault and so he should not be "written up" and punished. He reasoned that Alton had not shut off the gas and that the muffler on the vehicle must have gotten hot and ignited it. According to this interpretation, Alton was responsible for the release of the gas, but not its ignition. Not everyone accepted this explanation, though they did not mention alternatives. Big Daddy justified this view on the basis of a video he had once seen depicting a similar chain of events. As one mechanic said over lunch, adopting a mocking tone, "Yeah, 'cause everything you see in movies is real, right?" Some were bewildered, claiming that Big Daddy would get a reduced end-of-year bonus as a result of the incident. They joked that Alton must be a distant relative of Big Daddy's without our knowledge. Bob wrote up Alton for a past accident instead, some claimed in retaliation for being overruled.

Alton came back two-and-a-half weeks later, nervous and angry. He showed us the severity of his grisly injuries, criticizing the appointed doctor who had cleared him for work. He disputed the official version of events, claiming he had switched off the pump and that he should, therefore, be reimbursed for his medical costs. Skepticism about Alton's responsibility for the incident was largely due to the perception that he was accident prone and incompetent. I had not heard these claims until his accident, but afterward they circulated much more often. Some of his fellow operators insulted him, claiming that he was being phony and "ass-kissing" management to avoid getting fired. They mocked him for being unskilled and clumsy with the water wagon as well. For a time after the incident, people would bring it up when near the ash cells. While we moved an old leachate hose to the pond near where the accident took place, people said, half joking, that this is where "he doused himself." The tone of the mockery was typically playful and pitying rather than hateful. A round of intense mockery might conclude with someone saying, "But he's got the biggest heart in the world, he just has no common sense!" Bob mocked him in a similar way. After humiliating Alton for what he called "groveling," Bob waited until he left and added, "That Alton's funny, but he's a good guy."

Alton was to be mocked and pitied because he had proved himself hapless and incompetent. He had to face repeated humiliations at work, in addition to suffering through the pain of his recovery, as if the accident had shown him to be a pathetic fool. The man I knew was generous and kind and had overcome a struggle with addiction. Because safety is equated with responsibility (and every worker fears for his or her own safety), an accident can become essence: it can seem to reveal who you really are. The abstract labor power for which you are paid then becomes painfully concrete, leading to further injury and indignity.

TAKING SWEET OLD TIME

The social and corporeal difference marked by the phrase "smells like money" dissimulated other experiences of work discipline and conflict at the landfill, reducing them all to an individualized pursuit of money. Exchange value was not the dominant story in all work situations, or for every worker. In fact, some laborers claimed to have willfully chosen a job with less pay and fewer benefits and without union representation or overtime.

Mac and Timer had worked at Four Corners for more than a dozen years as laborers and had continually refused opportunities for promotion. While we worked together side-by-side one morning, they informed me that Bob had once offered to promote them both to operators, but they had quickly turned him down. This puzzled me, knowing as I did how precarious their finances often were, and I told them as much. Some weeks later, Mac and Timer brought the issue up once more, seemingly intent on giving me a more elaborate answer. We had just arrived at work and were discussing a young operator who was recently fired for being involved in his third accident on site. Others were saying that it was cruel of Bob to fire him so close to Christmas, during the long off-season for construction work and right after he had bought his first house. "See," Timer said, turning toward me. "*That* is why I don't wanna be an operator. I probably could be, but I don't want to deal with Bob's shit!" Ned represented a similar case. He had been one of Bob's boys originally but had gradually fallen out of favor. Though he was a shift manager, other operators began to tease him for being lazy and incompetent. Most laborers were fond of Ned and did not partake in such abuse. Mac echoed his older brother's claim: it was better to remain a laborer than to submit oneself to closer supervision and evaluation. He equated becoming an operator with working directly with Bob, which we only did rarely, when selected for special tasks.

Industrial labor is often characterized as if it were solely dominated by managerial time-discipline—that is, the rule of the punch clock. "Time is money," and according to this conception of work, any time on break or resting is considered time wasted. But time is not singular and indivisible; the multiple temporal rhythms of our lives are continually negotiated.[15] Time-discipline at work has different manifestations, depending on how regular, coordinated, and/or standardized labor activities are. For laborers at Four Corners and in related industries, time-discipline is highly regular, insofar as they work the same general hours and repeat similar tasks, but it is hard to standardize and coordinate. It was difficult for me to learn and adopt the multiple temporal rhythms of paper picking, which involved alternating between focused work, conversation, play, smoke breaks, and bathroom breaks. This was necessary to avoid "going at it too hard" and wearing yourself out.[16] It was always possible that the fluid slowing and quickening of labor could be interrupted by managerial authority and time-discipline, but this was rare and unpredictable. As laborers, we were more task-oriented and less governed by management. Operators were far more disciplined in their temporal rhythms, though not entirely, and new monitoring technologies were then being implemented that would keep track of compacting labor.

In other words, choosing not to be an operator was about being able to choose how hard to work and when. "Bob rides you," Mac would say. "He rides you and rides you until you can't take it no more." He and his brother liked to talk about people who had quit because of Bob pushing them too hard; one laborer, in particular, was being groomed for a promotion to mechanic until he lost his temper and challenged Bob to a fight. It was not simply that Mac and Timer were avoiding further responsibility by remaining laborers; they were avoiding the further indignity of having to constantly behave insincerely and act subservient as one of Bob's boys. Such freedom from supervision is a familiar ideal among North Americans. Out picking paper, not only the boss but also the impersonal forces of corporate bureaucracy and market dynamics seem distant. The pursuit of money is not the only source of transcendence, of escape from one's circumstances. Mac and Timer's shared desire to remain laborers may have been as much a result of feelings of inadequacy as resistance against Bob's authority. But it was not so very different from middle-class aspirations that operators have for their children—who, they hope, will choose something dignified and freeing.

The homes of the laborers Mac and Timer were also very different from those of their operator counterparts. Both brothers would wear their work clothes home—"home" consisting of two houses on landfill-owned property (across the road and within its boundaries, respectively). Mac, in particular, bemoaned his lack of separation from the landfill and regularly complained about the clouds of dust that passing vehicles would create while he tried to enjoy his days off. Having Mac's home on site, just within the boundaries of the perimeter fence, meant that we could take breaks and drink from his garden hose. It also created problems for him, particularly when the landfill began insisting on locking the perimeter fence, preventing people from stopping in and visiting and, Mac complained, making him feel like a prisoner. The freedom to receive and host guests, for Mac, was an important part of the open-ended temporality of being at home and sharing it with others. Opening and closing this kind of access according to the official business hours of the landfill seemed like a perversion.

Timer lived across the road and had related complaints, particularly concerning the regular use of his backyard to burn a fire to block odors. Timer would also wear his uniform and boots home, not bothering to change in the locker room with the rest of us. As I discuss in chapter 3, his leisure time at home was usually spent hanging out in his garage, tinkering with machines and doing yard work, and it made little sense for him to don new attire. The stuff accumulating in his garage, moreover, was not separated into things he'd bought, been given, or found at the landfill, although he remembered the social histories behind each item, especially if they represented a social debt. Like his stained, tanned, and weathered body, his garage did not help him transition into anonymous bourgeois domesticity—as in the homes of George and Dan—but represented a site of relative autonomy from both his job and his family, from Big Daddy and his Old Lady.

Mac and Timer were markedly different from many of their coworkers, not least because of their position as lower-paid and less skilled laborers. Even so, the anxieties they voiced betrayed some attachment to the middle-class ideal of autonomous houses, separate from places of work and the threat of corruption from mass waste. But such an escape from transience is only ever temporary for anyone.

THE ENDS OF VALUE

Aspirations of transcendence through value-accumulation have led to palpable tensions in the work and home lives of North Americans, beginning long before the housing bubble burst in 2007–8. My Midwestern coworkers (not to mention other friends, acquaintances, and fellow graduate students) invested in prominent value-forms with an expectation of overcoming their pasts and opening up possibilities for the future. Prior to the financial crisis, people living in houses they could not afford in the Detroit metropolitan area were known as "house poor"—a social category with which I was not previously familiar but have now encountered elsewhere. Their dwellings did not represent their existing financial circumstances but a future they aspired to achieve. Every virtual investment in a hoped-for future potentially runs aground of the actual world and its limitations. One may profess to smell the money even while sinking with a toxic asset.

David Pedersen argues that, in general, talk about money in capitalism drowns out other stories, "both what is necessary for its formation and also what may contradict its story or even lie relatively exterior to it" (2013: 17). Waste circulation is one such condition of possibility that is necessary for the formation of value, and yet it seems to lie on the outside, at a distance. Alternative accounts are always possible, however, as when a cherished investment in a house turns out to have been a toxic asset all along. Emblematic concretizations of American value like vehicles and houses may dissimulate the real accumulation of debts and an unstable future. The college educations that many of my coworkers fantasized about securing for their children offer similarly illusory transcendence. North Americans now routinely fret over the acquisition of seemingly useless college degrees that may fail to secure employment. It may be that the dirty work of landfilling is a safer proposition. As I write this, people increasingly speak of higher education as a new bubble with mounting, unpaid debts and bleak economic prospects for graduates. Even anthropologists debate the morality of graduate education in the midst of a dismal job market for PhDs. I myself thought seriously about taking a low-wage job during several precarious years on the academic job market. At the time, I regularly told my wife that I had worked at a dump before and wasn't above working somewhere supposedly unworthy of my graduate degree so that we could get by. I had learned how to smell the money.[17]

The transcendence that many seek, as a counterpart to aspirational middle-classness and the fantasy of self-creation, is also reliant upon waste circulation and disposal. Waste management mediates between value registers, amassing dead commodities elsewhere, thereby allowing production and consumption to be started anew before culminating in further disposal. Money is partly responsible for the mass waste of industrial societies: by acting as the universal equivalent, money allows commodification to run amok, resulting in "the vast mass of differentiated stuff that surpasses our capacity to appropriate it as culture" (Miller and Horst 2013: 6). But everything worthwhile eventually turns to junk, and even money eventually dies, at least in the form of fiat notes and coins, which ultimately must be disposed of. National treasuries must eventually remove bills and coins from circulation, whether to maintain the health of the economy or improve surveillance of criminal activities. One can find old coins that survived incineration in the ash piles of Four Corners, enough (Big Daddy would claim) to fill a coffee can in one trip across. Picking paper, one encounters the endless discarded documentation of credit card accounts, which is so easily mined by identity thieves. This is the condition of possibility for concrete uses of the universal equivalent and the pursuit of transcendent value.

Investing in the waste industry's accumulation of negative value could be seen as the meta-bubble that guarantees the rest. According to Bob, Four Corners is not a dump at all, but a piggy bank. "This is America," he said during one tour of the site. "We can do anything we put our minds to and someday we will find a purpose for all of this." Others share his fantasy and find purpose in transcending material limits. During my fieldwork, a former investment banker began a company with the goal of preserving bales of waste using patented shrink-wrap technology, so that they could be stored until a future use is found for them. "Ultimately we'll be depositing it in our own landfills or in facilities that will be converting that garbage into useful products: biofuels, electricity, heat," he claimed. "What you'd effectively create with these bale-fills would be energy reserves" (Gelfand 2007). Dreams about the untapped potential of waste are hardly new. Fellow French exiles, Victor Hugo and Pierre Leroux were vocal proponents of the *circulus,* a vision of a socialist utopia that would tap the sewers of Paris as a vital resource (Reid 1991). According to these visions, sewers and landfills will one day cease being mere depositories for valueless materials, and will instead become the ultimate stores of value.

Visions of the future aside, landfills are already sources of revenue. They are proof that even the negation of materials, their removal and burial, can be transformed into price, profit, and possibility. Even now, waste in North America smells like money while all other stores of value begin to show their wear and rot. But, as I've argued, it can only do so through limited abstractions of labor and waste, a partial story about human and material potential. How might such potentiality be more fully realized and released? That is the focus of the next chapter.

THREE

Going Shopping

ANYONE WHO OBSERVES WHAT PASSES through the great North American waste stream will eventually come across something they would not have thrown away. You stumble across a working computer, a perfectly good television antenna, or boxes of alcohol still in the bottles, and find yourself asking, "How could someone throw away *that?*" At many of the waste sites I have visited, it is commonly said that people will dispose of just about anything.

Since the late nineteenth century, the growing middle classes of wealthy capitalist societies have been encouraged to replace old commodities with new ones and find use for things that they had no need of before. American corporations, social reformers, and political elites supported these acts of possessive accumulation and dismissive disposal. At the same time, "materialism" was acquiring a pejorative meaning. Just as public elites were lamenting the disappearance of the frontier and the guiding principle of Manifest Destiny, there was growing concern that consumerism would corrupt American character and society.[1] More recently, various waste scholars have called into questions overzealous critiques of mass consumption and its purported wastefulness. They point out, in particular, that many households expend a great deal of effort in order to secure appropriate routes of disposal for unwanted possessions—including, but not limited to, turning them into rubbish.[2]

The problem is not whether modern consumers are really wasteful, but, rather, the moralizing common sense that insists we discuss consumption in these either/or terms. To shed some new light on what we discard and what this means for us, I propose an altogether different question: Why is it that so few North Americans scavenge other people's waste? By *scavenging* I mean the practice of recovering what would otherwise be disposed of, in order to reuse it. So much focus has gone into what consumers do or do not wantonly throw away that it is easy to forget that being a mass consumer—one who buys things in shops that are made somewhere else by someone else—is itself a decision. The understanding of choice in consumer societies is often limited to what we buy, not whether we do so; it therefore involves implicit assumptions about how people are related to things and to each other. The question that goes unasked (except by scavengers) is why anyone would choose to be a consumer at all, with so much perfectly good waste lying around. That is, why would anyone spend money on new goods when they could shop for free by touring the neighborhood landfill or diving into the Dumpsters behind the supermarket?

One answer is that scavenging may be too difficult or perilous. There's an ambiguous transfer of ownership involved in disposal—it's one thing to be invited onto someone's property for a yard sale, but it's quite another to pick through what they have placed on the curb in sealed containers. And where property is so routinely privatized, as in the United States, anxieties arise about who really owns discarded waste and at what precise moment claims to it have been abandoned. Many food and supermarket chains exploit this ambiguity by preventing access to their waste, and the regular and systematic collection and disposal of mass waste assists them in this. Go to your neighborhood supermarket, restaurant, or cafe and look for Dumpsters to scavenge from. You will probably discover that they are locked or otherwise inaccessible.

Scavenging food from industrial suppliers is considered transgressive because it interrupts established routes of material preservation, purchase, and disposal. It is not just that people *will not* scavenge; all over the world, government regulators, corporations, and waste industries prevent people from doing so. For example, we are generally denied access to Dumpsters that contain spoiled food from supermarkets (perhaps only slightly past their sell- or use-by dates), and only a select few are permitted access to the contents of garbage trucks, transfer stations, and landfills. At Four Corners, "salvaging"—as it was often called—was disallowed according to company policy, but this was not strictly enforced. Even those with access to mass waste are often formally discouraged from selling or reclaiming things they find in the waste—they are, after all, paid a wage to treat the material as valueless.[3]

Generally speaking, only those willing to challenge authority or bend the law can actually get at mass waste. But most North Americans are not likely to think that they are being denied access to a resource; they would consider scavenging as something done only by necessity, when one is homeless or trying to survive a zombie apocalypse. Because scavenging is perceived as degrading and dirty, it is assumed that no one would do it unless they had no other way to satisfy basic needs. That scavenging should appear this way makes sense if you are accustomed to capitalist exchange, in which transactions are based on purportedly free transfers of ownership mediated by money. For many North Americans, it might seem logical to assume that people search dumps for scraps to reuse and recycle simply because they lack the financial means to engage in normative acts of exchange and are desperate to survive.[4]

But what appears to be simple material necessity could just as easily be seen in terms of social possibility: when people discover and remake things, they do more than survive and can demonstrate what they themselves are worth—that is, that they have the skill to find and make use of something that another did not.[5] If, in normative capitalist exchange, people and things are subsumed by the abstract potential of money as universal equivalent, scavenging involves more concrete and serendipitous enactments of human and material potential. Mass consumption is more than a way to acquire goods; it is associated with the belief that buying things is preferable to finding or making them ourselves, just as acquiring things new is considered better than recovering them from someone else's garbage.

In this chapter, I consider the foreclosed possibility of scavenging in relation to the desire to buy and have things new. Mass consumption and mass wastage are generally understood to be related to one another as cause and effect, respectively. On one level, throwing something away leaves space for its replacement, which is what makes possible planned obsolescence, but the relationship between waste and consumption goes further. While it is true that mass consumption leads to mass waste as a byproduct, mass wasting is far older. Urban denizens were expelling waste en masse before industrial products were widely available to the masses. More to the point, without certain forms of wasting, mass production and consumption would not be possible. Most of the mass waste landfilled in North America is not a consequence of what individual consumers reject but is wasted prior to consumption, as a result of processes of extraction, production, exchange, and their regulation.

We are often told that mass consumers see commodities as merely disposable, when, in a sense, the opposite is true. According to the concept of planned obsolescence, and the moral critique of consumption associated with it, commodities are sold knowing that they will be discarded someday soon.[6] If the inevitability of process and change means that everything is on its way to becoming waste, planned obsolescence describes how commodities can embody transience in their very design. However, as part of the lived reality of North American life, consumption is equally premised on the durability of what we buy. And the mass consumer's decision to buy and sustain valued items means that they dispose of many other, secondary, things in the process. When shopping, they are ideally presented with a pure and unsullied commodity neatly swaddled in plastic, paper, and cardboard. I call these various, humble supports *bundling waste,* characterized by an ability to help other things last and display their durability to others. Before it is unceremoniously disposed of, bundling waste prepares the representational grounds for the commoditization of discrete and salable things. The form of the commodity is thereby reconstructed as the same, an object faithful to the corporate brand and thus separable from its singular creation and history. And if a commodity is highly valued, mass consumers may even invest in further material supports to preserve it over time. Anything we care for will expend waste if it is to last: electronic tablets and smart phones may go through many replacement screens; vehicles may go through many tires, brake pads, and engine components; dwellings may be rewired, their plumbing altered, walls torn down and erected, floors stripped and retiled, roofs repaired—and in each case, secondary components are discarded in order to maintain a more durable form in perpetuity, to continually reconstruct it as the same.

In fact, landfills are mostly filled with materials and not goods—that is, with the paper and plastic leftover from commodity care rather than the commodities themselves. These materials are clearly related to processes of commoditization and are expended to cover up the singular imperfections and unique histories of what we buy and own. In the same way that mass waste disposal creates (limited) opportunities for scavenging, it also facilitates the consumption of fresh commodities that appear to transcend the threats of process and change. Contrary to critics of mass consumption, North American moderns are primarily drawn to shop, to consume things new, by the *virtual permanence* of things for sale, and not by their actual transience. Buying new, again and again, not only eliminates the old but also conceals this very process of continuous renewal-through-transience (see Hawkins 2006: 129).

When people scavenge from mass-waste streams, they also circumvent the tacit norms of mass consumption, which keep the ideologically purified domains of home and market in dynamic tension. This two-sphere ideology purifies the realm of domestic relations and intimacy from that of cold rationality, work, and commodity exchange. While in actuality these opposing spheres intermingle and mutually constitute one another, I argue that landfilling and shopping maintain their virtual separation. Scavengers complicate these routes of transfer, though in doing so they do not defetishize the commodities they recover: that is, they do not necessarily use reclamation to reflect on the broader politics of production that lie beyond the surface appearance of market exchange. However, employees at Four Corners do *estrange* commodities from prevailing circuits of value creation and destruction. Lucky finds and clever repairs interrupt the dreamlike, phantasmic ways in which subjects and objects ordinarily relate to each other within consumer capitalism. A bra used as a flag to decorate gas wells, scrap metal used to supplement income, old and strange pornographic media and toys circulated for amusement—the irreducible potentiality of landfill contents, the sense that anything could be in the waste, provides person–thing encounters at Four Corners with a degree of ineluctable spontaneity and sensuousness.[7]

In discussing tinkering later in this chapter, I will turn the focus of analysis toward *shop*, not as a verb, but as a noun that refers to a privileged site of expertise (as in *shoptalk*). At Four Corners, the productive skills that fade from view in consumers (mere buyers of things) are brought into relief through the highly prized ability to rebuild and tinker with machines. In North America, the creative power of scavenging and tinkering with old commodities represents a source of creative transgression as well as masculine pride, pleasure, and sociability. For this reason, both workers and managers at Four Corners actively sought to acquire and repurpose things creatively. This allowed them to circumvent the formal predictability of the market—in the sense of both the constrained sale of commodities and the limited provision of a wage with which to purchase them. At the same time, irregular acquisition could allow them to circumvent the intimate politics of the home as the ideological obverse of the market, where commodities and wages are absorbed and displayed. Capturing and remaking estranged commodities can be a source of social relating, but they also make possible highly gendered acts of self-creation in productive tension with the ideological division of market from home. Put differently, it might promise a way to opt out of this forced choice.

North Americans need not scavenge to escape the safe and predetermined object–subject relations associated with shopping and landfilling. At the end of this chapter, I will return to the relationship between the refusal to scavenge and the purported decline of handwork or skillful tinkering. Just as scavenging estranges the ordinary routes of disposal by which transience is transferred elsewhere—into bundling waste and landfills—the way in which landfill workers embrace tinkering demonstrates why the use of objects cannot be conflated with a model of consumption as the destructive annihilation of goods. Practices of creative tinkering have not, in fact, disappeared but have been shuffled to new sites, such as computer, car, and home repair. While scavenging remains foreclosed from many people's experience, tinkering thus remains an exalted and highly gendered testament to bourgeois self-creation in contemporary North America. Just as we embrace the proliferation of disposable bundling waste as insurance against process and change, North Americans still seek out new and alternative ways of relating to things and each other.

Disposability sustains mass consumption by enabling the sociomaterial conditions of possibility for certain types of capitalist exchange, specifically shopping—that foundational practice linking the ideologically opposed, but mutually constituting, domains of home and market. In its very design, a sanitary landfill forecloses the possibility of reclaiming items, which was critical to its increasing popularity among early-twentieth-century waste modernizers. The scavenging I witnessed at Four Corners revealed the conflict between recovering items from waste and buying them new, or between the choices to scavenge or to consume. These examples also help demonstrate the relationship between the tacit refusal to scavenge, the normative desire to shop, and the irresistible rise of landfills over the past century.

THE CONSTRAINTS OF MARKET EXCHANGE

One day, while I picked paper with Eddy, Timer, and Mac, the latter two reminisced about when the waste used to be better. As was typical of conversations about the way the landfill had changed, their narrative related the worsening quality of the waste to worsening work conditions under Bob's management. When they had had regular use of a truck, they said, the operators working up top would radio and tell them when certain loads had arrived, to which they were given unlimited access. Without the ability to communicate with operators, travel at will, and haul away larger finds, laborers could only salvage what could blow in the wind, roll down the slope, or surface at the Citizen's Ramp.

To illustrate their point, Mac and Timer explained that Four Corners once had contracts with a food company that would throw out whole cases of snack cakes in bulk. Whenever those loads arrived, the operators at the time would radio their laborer's truck so they could grab free food. This story led to a familiar exchange:

EDDY: [responding with a grimace] There's a reason they were thrown out!

MAC: [unaffected by Eddy's display of repulsion] They were still in the plastic and everything.

MAC AND TIMER: And they were damn good too.

The debate was not settled, but it did make clear where people tend to differ on scavenging. No one at Four Corners would unthinkingly eat something they found in the garbage, but things still in cases and packaging were in a gray zone of ambiguous quality. Mac and Timer reasoned that plastic packaging protected the cakes from being contaminated during their transfer from the factory to the landfill, despite mixing with other garbage en route. Generally speaking, whenever retrieving something from the mass waste stream, it makes sense to scan it for the copresence of something undesirable, a trace of its fellow travelers and a threat to its discrete form. By contrast, Eddy voiced the common, though generally unstated, position of the mass consumer: there are norms for quality control that corporations and government agencies have put in place on our behalf. This can be further premised on the self-interested and competitive character of market exchanges—if a company *could* sell something, they would—although it might also be represented as a form of care. Either way, Eddy was using this logic to question the recovery of any item that came directly from producers and was not first bought and discarded by a consumer.

It is not hard to challenge the consumption of mass commodities of any sort in the way Eddy did. Similar arguments concerning production and preparation linger over the industrialization of food as a whole—one need only think of the common North American joke that it is better not to know how hot dogs are really made. Concerns about the specific histories of consumables reveal a basic problem for any commodity transaction: how do you make it so that a good is represented as something suprasensible rather than sensuous, a formal value commensurate with its money-price and equivalent to every identical one you could buy, have bought in the past, or might buy in the future? The difficulty of reconstructed sameness arises especially when it comes to things we physically consume, but the argument is the same whatever the commodity.[8]

The greater the scale of mass production, the harder it is to account for a particular good's singular journey into our possession, which raises the specter of poor or inconsistent quality. The debate I witnessed reveals some of the hidden material operations that we rely upon, that are required in order for the rites of mass consumption to take place. These inform our judgments whether or not those snack cakes are thought safe to eat. Both Mac's hypothesis as to their continuing value and Eddy's calling this into question presume processes of standardizing, packaging, labeling, and regulating the goods we buy. For this system to mediate material transfers from sites of commodity production to the marketplace and the home, things of insufficient or questionable quality must be readily disposable.[9]

In part, Timer and Mac were reacting to what can be a matter of considerable frustration for scavengers. Most waste encountered is not found still in the package, as if newly bought from a shop. When this happens, it feels much more like you have somehow cheated the limits of your wage and the marketplace.

It is widely recognized that packaging has an important role in the history of mass waste. By some estimates, packaging regularly accounts for more than a third of waste generated in the United States, and this has made it a primary target for most recycling initiatives. But what of packaging's role in mass consumption? This can be understood as a means of bundling particular forms in such a way that representations of them are stabilized, making them more readily exchangeable on the market as discrete commodities. Bundling serves to hold together and preserve items for sale so that retailers, consumers, and regulators can count on or enact their suitability for purchase.[10] Bundling makes would-be consumers confident that items represented as formally identical in brand and quantity are also substantially identical in quality. Acting as both virtual and actual protectors of goods, bundling materials assure that commodities have not suffered from tampering or damage on their way into the possession of consumers.

The rise of Euro-American recycling initiatives in the late twentieth century have brought packaging into public consciousness, but only as a remainder of consumption. Considerably less attention has been given to the inverse relationship, whereby packaging shapes consumption. Partly this is because packaging is forgettable by design. Not valuable in and of itself, but useful only insofar as it bestows enduring value on something else, packaging facilitates commoditization. There is a popular North American joke that children who receive a gift derive more enjoyment from playing with the box it came in. It is funny precisely because it violates the general rule that bundling material is designed for disposal. In this sense, packaging material is best thought of as *bundling waste*—because bags, boxes, plastics, and paper are merely temporary envelopes, meant to encase and preserve a commodity on its voyage from production to consumption, made for easy discard.

Things that are bundled and preserved do not always become commodities. The primary tool used by paper pickers, besides gloves, boots, and uniforms, was a roll of clear plastic bags. We would take several of these with us when we went out to pick, going through as many as four or five in a day at times. A garbage bag is designed to be simultaneously a functional container for waste and a kind of waste itself, readily disposable for both paper pickers and anyone removing garbage from their household. And yet this still affords commoditization, not of the waste exactly, but of the labor power of garbage collectors and landfill workers—by encasing garbage in a discrete bundle, garbage bags allow for the circulation of mass waste and its translation into profits for the waste industry.

Whether preserving something valuable or something disposable, bundling waste itself is inherently disposable. It may be recycled (and, indeed, recycling containers are typically custom-built for plastic bottles, cardboard, tin cans, and the like), but its disposability, by whatever means, is assured by its subsidiary and supportive role in circuits of value creation and destruction. Every commodity is bundled for sale, and its bundling does not contribute to its value so much as prepare the way for it to be recognized as such—that is, prepare the representational ground for our encounter with the thing in the guise of a commodity for sale. Interesting exceptions are those objects whose own form contributes to self-preservation (e.g., a car or bananas). Even thus naturally packaged, buyers still look for traces of decay, including chipped paint and a rusted exterior, or graying of the peel. But, in general, bundling waste serves to dematerialize exchanges by promising a relative detachment of the commodity from normal spoilage. A soda can with a punctured tab may have gone flat, groceries carried in a flimsy paper bag may spill out the bottom, a television screen delivered with insufficient packaging peanuts or bubble wrap may crack. By protecting commodities from process and change, bundling waste facilitates reconstructed sameness and sustains the market in virtual, suprasensible values.[11]

Bundling waste helps sustain the quality of commodities for sale, limiting the possibilities of decay and imperfection. This was perhaps never more important at Four Corners than when carrying around lunch. Some used lunch pails, but for those who patronized the Snack Shack or Roach Coach, it was vital for food to be enclosed in styrofoam containers and plastic bags as it was transported (usually by vehicle) to eating areas. Sometimes people would eat in the cabs of their machines up top, eat right at the Roach Coach, or leave the landfill entirely during their lunch hour, but often operators and laborers returned to the maintenance building to eat beside mechanics and, very occasionally, managers. I often purchased lunch at the Roach Coach, sometimes on credit, and on my walk back to the break room would carefully conceal my lunch in its container, folding the opening of the plastic bag underneath so that no dust or other pollutants could get inside. The meal I paid for was thus ideally preserved, though sometimes I would open the container to find that my food had been jostled along the way, resulting in a mess I had to reassemble before disassembling through (literal) consumption.

When I offered to volunteer at the Roach Coach for a short time, helping Bob's sister Tanya prepare meals and handle customers, it was clear that the bundling of commodities was as ritualized a procedure as preparing the food itself, carefully done to ensure quality. On a warm spring day, Tanya brought an electric grill, her great-grandmother's, and we set it up behind the food truck to precook burger patties and barbecue chicken breasts. After the meat was cooked, we left it resting in "burger juice" in a Tupperware container on top of the steamer, ready to distribute in the lunchtime rush. My field notes show the importance of bundling food for both preparation and sale:

> [We] lay out slices of bun, without condiments [onions, ketchup, mustard, and relish] because we have laid them out . . . on the ledge on the other side of the ordering window. You have already placed a "bubba burger" [named after Bob] . . . into a white [microwavable] container, covered in grease and melted cheese, that is reused to preserve and save product. You microwave the bubba burger for twenty seconds, if necessary with cheese. [After it is removed from the microwave, the burger is placed on a bun] in a new white container with a sheet of wax paper underneath. Or in a white styrofoam container with two sides for an extra dollar you get potato salad or fries or tater tots.

What was paid for at the Roach Coach was the food, the use value of which was its taste and convenient location. As I wrote in my notes at the time,

All the prices have been figured out by Tanya according to how much she sells per item from the total cost of the stuff she buys gross: she shows me a tiny piece of paper where she's scribbled [translations from price] into her cost per order.

Tanya invested so much in those containers and bags for the same reason that the frozen food she bought and prepared came in sealed and unmolested containers of cardboard and plastic: to help facilitate the food's transfer from preparation to sale to use. Everything happens as if bundling waste absorbs the qualities of transience that threaten commodities for sale. This resembles ritual sacrifice, whereby a relationship with the transcendent or divine can be established only by expending something more humble and transient. In the sacrificial economy of ordinary consumption, transience is channeled into the disposable sheaths that reconstruct commodities as the same and then are quickly discarded—and, in North America, usually landfilled.

The rise of refrigeration and plastics increased the availability and dependability of products and made possible new relations of trust and reliance between consumers and the food market. Occasional outbreaks of *E. coli* in lettuce or of mad cow disease in hamburger may elicit temporary shock and panic, but this hardly threatens overall confidence in the food industry. Influenced by the contaminating properties of mass waste, Eddy's distaste for landfilled snack cakes demonstrates a general unease with thrown-away things of uncertain origin. For all we knew, they were thrown away because they were spoiled, and that does happen. In fact, modern grocery shopping is highly dependent on a series of regulatory interventions that require the prominent display of sell-by, use-by, best-by, and expiration dates to the potential consumer. These physical labels, like the bundling waste that they are written upon, are themselves temporary and disposable. They may contribute to the uses to which foodstuffs are put, influencing decisions of when to cook or throw away, but in all cases, they mediate the risk and uncertainty associated with becoming a mass *consumer,* in both senses of the term. The result is a benign vision of market transactions, cast as a process of gentle competition rather than exploitation.

Consumers themselves may remain indifferent to these quality-control standards. They may also treat the preserved value of comestibles as something to be sacrificed for things and relations of greater import. I recall accompanying Timer to the convenience store on his son's birthday, where he purchased an inordinate amount of sliced meat given the number of people in attendance—which included myself, his two sons, a family friend, and an aunt—few of whom actually came to eat. The ham itself was protected in a plastic bag, meant for opening and resealing by design, thereby steadily reducing its quality over time. The point of the gesture, however, was precisely to buy food in excess of what was needed, thereby demonstrating to all how important the occasion was and how hospitable he could be (the excess beer he purchased was another sign of Timer's intentions, though this would be readily consumed long before the meat began to go bad).

And yet the politics of spoilage operate behind the scenes of consumption, whether or not people are aware of them or explicitly opt in for the protection of their financial and bodily health. The possibility of spoilage has so shaped the political economy of food in the United States that it was in response to the hazards of transporting grain by boat and rail that the first futures exchanges opened in Chicago in the 1860s.[12] Additional bundling waste includes, arguably, the receipt itself, which has played a further role in shaping the financialization of everyday consumption and the microhistories of accounting. The receipt is often disposed of as well, but in modern capitalist economies it represents a tiny surplus gift to the consumer, a token that provides further assurance as to the quality of the product and the reliability of the seller.[13]

Bundling waste may not always be so disposable or forgettable. In cases of identity theft, for example, exchange-facilitating items like receipts or bank statements can be scavenged as a source of valuable information for illicit gain. I never heard of identity theft taking place at Four Corners, though new concerns about it were emerging during my fieldwork, with the growth of social media and digitized finances. Eddy, always the aspiring criminal, once claimed excitedly that he would try to use an old checkbook someone had discarded to write bad checks, though this proved impossible given that no pages remained blank. On another occasion, Eddy found a 1978 license plate in the garage of his new house and talked about how he might be able to use it for his own car, so that it could not be traced to him. I don't think Eddy would have gone through with either of these acts, but such illicit fantasies are made possible through the routine disposal of a wide range of humble objects whose purpose is to facilitate or account for the movement of commodities.

HOME AND MARKET

Until one tries to circumvent normal routes of material transfer through scavenging, it is easy to miss just how deeply bundling waste is caught up in today's consumer societies. Scavenging possesses transgressive qualities in relation to the tacit norms of mass consumption. Fundamental to these rules, written and unwritten, is the assumed divide between the spheres of home and market. If the practices of working and shopping mediate the shifting relations between these two virtual domains, scavenging disrupts them, estranging commodities from the ordinary routes of transfer though which they are connected and kept separate.

The Two-Sphere Ideology

As discussed in chapters 1 and 2, most of my former coworkers were heavily invested in a material and moral distinction that divided life in the landfill's orbit from life at home. The distinction between public and private life is part of a historically gendered and racialized two-sphere ideology that imagines these domains as wholly and naturally separate from one another. The two-sphere ideology is an ideal model and one, I would argue, that many of my coworkers would have instantly recognized and that they actively lived by, though they did not have a name for it themselves. For the purposes of this chapter, I will gloss the two spheres purified by this ideological model as "home" and "market."[14]

In North America, over the past century, markets have tended to be imagined as places of profane competition and material production, of commoditized relations whereby money is exchanged for things and labor. The home, by contrast, has tended to represent a place of sacred communion and social reproduction where things and people are nurtured and cared for without expectation of a return, a place of free gifts and hospitality. These ideological associations reflect a broader imaginary whereby certain possessions of great personal importance (family heirlooms) are defined by their lack of circulation, or inalienability, while other things (money, wage labor) are defined by the ease with which they circulate, or their alienability.[15] In this sense, the disposal of things helps provide clear distinctions between possessions that are valuable and durable—meant to grow together with people—and those that are worthless and transient.

The problem with the two-sphere ideology, in the guise of home and market, is that it misrepresents the many ways in which these social domains are continuous and have always been, because they share similar attributes and mutually constitute one another. My informants would readily recognize this contradiction. In southeastern Michigan, kin effects and relations are not absent from workplaces as if belonging to a distinct realm; the two regularly interweave. This is most obvious in the way people talk about the omnipresent auto industry and the highly coveted union jobs that increasingly require family connections to acquire. Even the local name for the auto industry—"Fords"—reflects the use of kinship, personalizing the impersonal corporation as if it were yet another local family. The interpenetration of work and family was also apparent at Four Corners. Although their jobs were not nearly as valued as union jobs at Ford, several employees owed their positions to kin relationships. Todd was the brother-in-law of Corey; Timer was brother to Mac; the mechanic Zack was Bob's nephew; and the on-site eatery, the Snack Shack, was operated first by Bob's wife Debbie and then by his sister Tanya. Henrietta, the lead administrator at the landfill's sales building (called "4000"), had helped get her son Eddy his position as a laborer; her "honorary niece" Andrea a job at the scale-house; and Andrea's mother, Penny, a clerical job at 4000. Bob and Big Daddy both liked to think of themselves as father figures for Eddy, who had lost his father at a young age. Debbie's words were not entirely metaphorical, then, when she described the landfill employees to me as "one big family." Those without relations on site had family ties to the area and the land. One of the newer operators happened to be a good friend of Timer's estranged daughter, who had become a stripper in the neighboring town. Ned, whose story I discuss in more detail in chapter 4, grew up farming the plat that the landfill now owns.

Critiques of the presumed separation of home and market do not always explain how these domains come to seem separate in spite of their resemblances and interconnections. Marxian analyses have described the coproduction of market and home in a way that accounts for the concealment of their unity, focusing on the selective appearances of people and the things transacted between them, and the interests these serve. This occurs through a gendered division of labor, which enables the social and ideological reproduction of workers to sustain capital—and wages to sustain the family.[16] But families not only produce workers for capitalist industry and live off their wages; they also consume industrial products in a way that makes home appear to be something other than a simple domain of social reproduction; as touched on in chapter 2, the aspirational middle-class home is simultaneously a dwelling for consumers and an object of consumption.

This historical significance of shopping both expresses the ideal model of the two-sphere ideology and exemplifies its contradictions. With this social practice, the products of the symbolically masculine domain of work—money and commodities—become domesticated as part of the symbolically feminine expenditure. Shopping also evokes sacrifice, insofar as it serves as a site of conversion across seemingly opposed domains, including male/female, work/leisure, home/market.[17] The hybridity of shopping partly arises from its mix of ideologically feminine care with the distinct rationales associated with the normative, masculine idea of the market. In many North American marketplaces, and in meaningful contrast to the affects and relatedness associated with home, rational calculation and self-interested behavior are seen to be appropriate and are materially facilitated as such.[18] As with a high priest engaged in a sacrificial ritual, who carefully compartmentalizes sacred domains from profane, the celebration of cold, self-interested calculation while shopping is held at a remove from the warmth and selflessness of home. With the aid of coupons and catalogs (yet another kind of bundling waste), the potential conflict between selfish shopper and selfless provider is kept in proper balance, maintaining the presumed opposition between these sites and selves. The hybrid practice of shopping enables us to dwell in two spheres at once—to find the perfect gift for a loved one at a competitive price.

The boundary-crossing act of shopping represents the normative material traffic between home and market. As the example of bundling waste shows, shopping is also a clear example of a material arrangement made possible by the subtraction of mass waste. If the two-sphere ideology envisions an ideal separation between home (the site of social relations and reproduction) and market (the site of wage labor and commodity exchange), then we can complicate this diagram by inserting the actual practices that mediate the relationship between these virtual domains (see figure 4).

As explained in chapter 2, landfills turn our waste into money by representing diverse waste streams through standard measures of volume and weight—which, in turn, can be translated into available air space and profit. We can now add that, in so doing, they help establish a profane and amoral market in competitive disposal, the political consequences of which I discuss in chapters 4 and 5. The commoditization of waste services, in turn, provides a condition of possibility for the bundled commoditization and calculative consumption of many other things, which lends actual support to the virtual separation between sacred home and profane workplace.

Acts of scavenging, by contrast, serve to disrupt idealized relations between landfill, home, and market. Attending to specific acts of scavenging offers an opportunity to further delve into the means by which mass wasting sustains the ideological connection-in-separation of home and market.

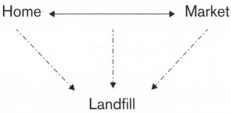

Home / Market
(Virtual separation presumed by *two-sphere ideology*)

Home ←——————————→ Market
(Virtual separation sustained by *actual* flow of commodities—goods, labor, money—between them)

Home ←——————————→ Market

Landfill
(Actual flow of newly purchased commodities sustained by *virtual* absence of wastes)

FIGURE 4. The relationship between home and market.

Gendered Scavenging

Scavenging waste was also called "shopping" at Four Corners. Similar to the claim that waste smells like money, discussed in chapter 2, this was considered funny precisely because shopping at a landfill and at a store are very different ways of acquiring goods. Scavenging deviates from the routine and ideologically gendered movement of commodities between sellers and buyers.

Bart, laborer turned operator, told me about the time he brought a can of coffee home to his wife that he had found in one of the waste loads. He thought she would be pleased, but instead the can went right back to Four Corners not long after. The problem with bringing it home was not so much that the grounds were spoiled, but that his wife was uncomfortable with the bundling—there was a dent in the side of the tin. Failing to secure her confidence in the continuing sanctity of its form, it turned out to be unusable trash after all. This story of failed scavenging is gendered in multiple ways. Bart's wife is represented as a barrier to scavenging and, more generally, as the one who decides what consumable commodities may enter their home from the outside and how. In spaces ideologically represented as feminine, like many North American kitchens, proper bundling indexes not only product quality, but effective nurturing through consumption. Insofar as caring for one's family means, in part, ensuring that they are not being poisoned, discarding spoiled or unwanted food is an important complement to shopping.

Bundling waste can secure exchange from risks of contamination and corporate liability lawsuits only with the ready availability of cheap and convenient mass disposal. And the importance of landfills to processes of commoditization can be measured through their steady accumulation of both packaging and bespoiled materials. This is not only to meet the demand for bundling waste; it is also to permit secure routes of disposal for accidents of mass production—inadvertently spoiled, damaged, or otherwise risky commodities. By recovering such material excesses, or "write-offs," scavenging potentially challenges the mass disposal that underwrites the stability of market exchange as well as domestic nurturance.

As discussed in the Introduction, in the past, loads from local party stores occasionally had to be dumped at Four Corners because of smoke damage. Such waste loads, particularly when they included alcohol, came to the landfill escorted by government agents who had to guarantee that the items were properly disposed of before leaving, in order to make sure that the alcohol was not illegally acquired and potentially resold. A few operators were fond of remembering how easily they fooled the armed ATF agents who watched them cover the skids with a thin layer of waste. After the agents had gone, I am told, they scraped the garbage off and dug out and divided it among themselves, filling their garages—not their kitchens—with boxes and boxes of liquor. Both the operators' desire for the alcohol and the strategy by which they recovered it hinged on perspectives toward bundling: either cases and containers preserve the drinks from the damage of smoke and landfilled waste or they do not. How one comes down on the matter of bundling marks not only the difference between good and bad taste, furthermore, but also that between lawfulness and illegality. The reason the alcohol story has been repeated so often is that the spectacular find is made that much more significant by the simultaneous violation of different barriers and rules of conduct—governmental, managerial, and bodily—and the betrayal of the entire enterprise of mass disposal that sustains ordinary commodity transactions.

Acquiring illicit alcohol also reflects some of the ideologies associated with the hybrid practice of shopping, even while circumventing normal routes of consumption. In either practice, there is a value placed on getting deals, such that "it was a steal" and one acquires "something for nothing." Quality controls create standards by which commodities are demoted to lesser spheres of value. Buying something slightly damaged on sale is similar in this regard to scavenging, insofar as both reflect a consumerist ethic of wise competition in the marketplace. The expression "fire sale" has become a common way to indicate an event where goods can be acquired at bargain prices, but it no longer refers specifically to fire-damaged goods. Instead, low prices are dictated by market competition and a need to eliminate inventory, rather than a desire to unload commodities of dubious quality. Furthermore, it is telling that in keeping this excess alcohol—in excess of normal consumption twice over, as an outlier scavenged rather than bought and as a luxury desired rather than needed—landfill workers tended to load up their garages, which are historically gendered in North America as masculine sites of tinkering, storage, and display. Home is not a singularly gendered dwelling, and materials from the landfill could enter this domain if an appropriate route of transfer were available.

In fact, many of the stories of successful scavenging I heard involved gendered subversion, which does not mean that they subverted gender norms. Like Bart's failed attempt, scavenging stories often involve possessions associated with masculine enjoyment, particularly machines or machine parts and pornography. Timer was often on the lookout for the former. The area in and around his garage was a way station for recovered items from the landfill. On one occasion, he took great umbrage when Eddy prevented him from acquiring a TV antenna that they had spotted while cleaning the Citizen's Ramp. Timer had been looking for "rabbit ears" to go with a used TV he had acquired from an operator and complained about the offense all day, repeating in his usual style that he "almost said something" to Eddy, an expression that was meant to indicate not impotence but barely restrained rage. Other times, the roles were reversed. Eddy once complained when he saw a boat being dumped up top, adding that he wanted one very badly and other people were too lazy to fix up the ones they had. Timer responded proudly that he had once thrown a boat away that he had acquired from the landfill and that he no longer had a use for. In both these cases, for laborers of different ages and backgrounds, the landfill represents a means of accessing luxury commodities for the aspirational middle-class man.

Scavenging could involve patrimonial pleasure as well. Timer once picked a trampoline from the garbage solely for the use of his children. He was proud of being able to provide for them in this way. "Kids used to love the thing," he remembered, though it eventually disappeared after a particularly bad windstorm. It was the same for Ned, who grew up farming the land that would become Four Corners. He was proud of the fact that he had recovered a large wooden beam from his grandfather's barn before it was demolished by the landfill to make way for future expansion. This was transported to his new home to the south and helped establish a material foundation for his new barn, thereby creating a direct link between his family homes as an adult and child. Here, too, scavenging offers gendered opportunities for social reproduction and care that ordinary shopping and ordinary wages cannot.

Scavenging stories could also represent more transgressive masculine pleasure (as in the case of the recovered alcohol). Items in this category did not have to leave the landfill to serve this purpose. On some occasions it was best that they did not, as when Todd discovered a bike by the Citizen's Ramp. With some free time on our hands, he, Eddy, and I tried to "take it over jumps" in a conscious reenactment of boyish stunts of daring familiar to all three of us (see chapter 2 on competence and masculinity). Todd offered to pay us five dollars to accomplish the feat, but we both "chickened out" (a wise decision given that the brakes didn't work, the gears didn't shift, and the chain would slide off and interfere with the pedals when it was in motion). Similarly, with the recovery of illicit material—like pornography—the expectation was that the find might not remain on site, but nor would it be shared in the domestic sphere, even as stories of its recovery were shared among coworkers. One of our assistant managers once discovered boxes of pornographic videos while digging up an old portion of the landfill to work on the gas field. He and a contracted worker from another company reportedly took some of the videos home with them. It did not matter that the boxes were faded and crumbling, because such material was not meant for public display and the deterioration of the outer bundling would prove no obstacle to private use.

Typically, workers would recover pornographic and sexual items when their circulation could be the source of humor and mockery. Since the material would likely be tossed again anyhow, it was sometimes better if the recovered object showed signs of its transient decay. Like most forms of sexual discourse between men at Four Corners, this was typically delivered in a homophobic and/or heteronormative fashion. During one lunch break, I couldn't identify a small, dusty black object on the break-room table. A visiting employee from CAT labeled this a "butt plug" with an amused cackle, suggesting to the room that he had left it there. However, no one took credit for the nonfiction book about the life of Jesus that had also been recovered that same day and intentionally placed nearby. The serendipity of finds at landfills creates opportunities for humorous surprise and juxtaposition, if they are properly taken advantage of. Eddy and I would regularly joke about the possibility of scavenging sexually graphic material and putting it in Todd's locker. I was presented with this opportunity when I discovered a book of pornographic stories focused on incest and found Todd's truck unoccupied. I purposely left it there, planning to tell him that I was the culprit, to complete the joke. He deduced that on his own, however, and complimented me on the gag.

I couldn't verify whether gendered and sexualized finds were more commonly circulated and kept than typical goods, but they certainly seemed to make for better scavenging stories. This is partly derived from the way in which scavenging estranges materials from the normal circuits of transmission that delicately link the ideologically gendered and sexualized spheres of home and market, private and public. Next I will consider the ways in which scavenging offers opportunities to opt out of the forced choice of this ideal two-sphere divide.

ILLICIT SHOPPING

Within capitalism, commodities are typically presented and perceived in a phantasmic way, usually described as "commodity fetishism." This is made possible by the separation of home from market, which conceals the social history of an artifact's creation from the consumer, even as it is advertised, displayed, and acquired as a means to satisfy their desires (Graeber 2001: 65). If shopping joins home and market together and allows fetishized money and goods to flow between them, then recovering items for free, whether through scavenging or minor theft, is distinctly different. Many things would come and go at Four Corners, and as in most workplaces, a supplemental circulation of stuff existed, subsidiary to the operation of the business itself. At the landfill's sales building, 4000, this might include staplers, paper, and other office equipment, but many other materials become available in such a complex operation. On one occasion, Bob was "bitching" at the youngest mechanic, Zack, for taking company vehicles for personal use. Later, Zack and Jerry criticized Bob's hypocrisy, claiming that he took gas, filters, and other items from the maintenance building for his personal truck. If recovering items from the workplace acts as a supplement to one's wage, these items, like wages, are not evenly distributed to all.

Illicitly scavenged things are not defetishized per se—the social history that went into their production is not made visible, and a commodity may still seem desirable in and of itself, rather than a product of alienating and potentially exploitative social relations. At the same time, illicit, nonmarket forms of acquisition can be said to estrange commodities from their ordinary paths of circulation. When things are acquired in this way, they circumvent the ordinary (though phantasmic) movement of commodities from production to use. Because they are acquired differently, they can seem radically opposed to the temporality and banality of ordinary shopping.

Serendipity

One day, I came to work to discover that there was a new refrigerator in the break room. It had been sold to Big Daddy by an operator's sister at a low price. Because it was a disused item that someone had clearly pulled out of storage, we had to take steps to ensure its durability—all the more important given its function as a bundling commodity that preserves the quality of other commodities. Big Daddy instructed me to clean the refrigerator out and I spent hours doing so, using special "409" cleaner on the inside and an air gun to blow out the coils underneath. Partaking in the recovery of this replacement, cleaning away the remnants of its past to make it a lasting object, made me curious about its future as well as that of the fridge it had replaced. But when I tried to ask people about the fate of the old fridge, curious whether it had been dumped up top, they would immediately respond, "Take it!" My inquiry into an object's biography was mistaken for acquisitive desire.

The only person I could strike up a conversation with about refrigerator history was Mark, the oldest laborer and on-site janitor. Once again, the main characters in the story he told me were the valuable commodities themselves, not the people who gave them form and meaning. Mark did not care for the newest fridge, nor the one that Big Daddy had just replaced, but the one before that—the first of three that had occupied the breakroom over the years—had been a forty-year-old classic. It was old, Mark told me, "but it kept things *cool!*" He had wanted that oldest fridge, he told me, but it had been dumped up top before he could stake his claim. While fetishistic in some respects, Mark's singular desire for *that fridge* and his memory of losing his one opportunity to have *it* is also unlike the normal procedure of shopping at a store or online, where any similar model can provide an identical substitute since it is reconstructed to provide the same experience. While there is certainly an element of serendipity to conventional shopping in person or online (i.e., finding just the right X to satisfy our desires or someone else's), the randomness of what one finds in municipal solid waste is far greater.

FIGURE 5. Mac picking paper, with a serendipitous find.

There is an element of uncanny serendipity involved in acquiring estranged commodities, as when the exact thing one needs or desires just happens to arrive in one of the waste loads. One day while picking, I found a strange remnant from my childhood. According to my field notes:

> [While picking the slope] I find a warped, but plainly visible Star Wars toy. Han Solo in Carbonite. I keep it and Eddy sees it [at the end of the day, resting in my locker]. He responded, half-jokingly, "You should put that in your book: 'Garbage isn't all crap, it's personal memorabilia!'"

Eddy was right. Holding that object, I could distinctly remember burying Star Wars figurines in the sandbox of my childhood home, my own imaginary Sarlacc and proto-landfill. In such instances, the phantasmic quality of commodity exchange becomes further distorted, as if what we value seeks us out, rather than vice versa. This quality of person–thing relations needn't be fetishistic in a Marxian sense. At Four Corners, the serendipity of scavenging was more commonly drawn upon, as in Eddy's crack at my expense, to demonstrate reflexive distance from the fetishized character of shopping. While we picked the slopes or cleaned the Citizen's Ramp together, Mac would occasionally hold up random objects, repeating in a flat and mocking monotone,

"Do you want this?" Not just anything would satisfy the joke, only objects that stood out in some way: an old sign that said "Greetings from Kansas" or a t-shirt with shark jaws on it that read "Bite me." While mocking the ability of completely random finds to satisfy paper pickers, the "Do you want this?" gag is therefore also premised on mass waste serendipitously providing interesting and unexpected items to reject.

Part of the enjoyment of scavenging at Four Corners, whether to acquire something permanently or for brief amusement, is that it distracts from the temporal rhythms of regular wage labor—rather than picking or cleaning productively, one has to take the time to glean individual treasures from the rubbish and evaluate them. Scavenging thus circumvents both the marketplace in commoditized goods and commoditized waste labor simultaneously. Moreover, the work of estrangement is clearest when an object can be acquired that not only sustains work relationships or challenges work hierarchies, but allows one to seemingly opt out of the home/market divide altogether, not only because it does not cost a wage and is not bought at a store, but because it may elude one's social responsibilities or subvert existing norms and laws.

People nervous about scavenging might still delight in seeking out luxury goods through extralegal means. During my first few months at Four Corners, Eddy regularly advertised to our coworkers that he could attain items for them for free or at an extreme discount. Jerry, the senior mechanic known for being anxious about filth and pollution, had no qualms about goading Eddy into getting him an expensive golf club. By this time, Eddy had already managed to acquire PlayStation 2 gaming consoles for Todd and another operator and had provided Zack with an XM Radio. Getting the golf club would require direct shoplifting, he told me, and while this did not bother him particularly, he was getting anxious about satisfying everyone's demand and continually having to prove himself in this way. No one inquired how this was possible; everyone likely assumed that Eddy was using illegal means to acquire the goods.

Illicitly recovered or scavenged items, unlike those bought from a store, fail to provide their new owners with a complete social history. They typically fall into one's possession without the bundling waste (boxes, plastic wrap) and virtual documentation (receipts) that would connect them more directly to a retailer and producer. The risks that come with undocumented goods— much like those associated with undocumented workers—also come with the extralegal advantages of acquisition on the cheap. Once again, the items

Eddy was asked to recover were often masculinized luxury goods like video games or sports equipment. These were items that workers could usually afford, but which they typically had to do without as part of a domestic compromise for the sake of the household.

Estranged Money

One day, while picking the access road leading to the open face, Eddy and I passed by a large hunk of metallic coils half erupted from the ground in a tangled heap. The market in scrap metal is among the largest economies of recycling in the world. Given that southeastern Michigan has historically been rife with old and disused automobiles, there were many regional scrap dealers involved in this global trade. One such site existed just down the road from Four Corners. Eddy was uninterested in what the copper coils might once have been, and instead talked about how easy it would be to get money in exchange if only we could pry them from the ground. This seemed like an impossible task, and he admitted that typically it takes company equipment to do so. This was not always worth the risk, however. If Big Daddy were involved, we might be compensated for the find, but Eddy claimed that Bob would try to keep the cash for himself. He said that when he and Todd had previously discovered a deposit of scrap copper elsewhere on site, Bob would have been the sole beneficiary if Big Daddy had not intervened and rewarded them with fifty dollars each for their efforts.

Bob's acquisitiveness was not described as purely selfish, either by himself or his employees, but openly attributed to his peculiar (though familiar) household division of power and supposedly emasculating marriage. Bob needed the money, Eddy and others often repeated, because his "old lady" would not let him have any. Bob's marital squabbles were often blamed for his behavior at the landfill—although not everyone excused him for it, sometimes pointing out that they had difficult spouses too and they did not take it out on other people. Either way, this had been part of Bob's scavenging ethos long before I worked there (when he harvested the aluminum that came from demolishing Ned's ancestral home) and continued after I quit (when he recovered the copper from the old sales building, 4000, before the rest was added to the landfill). Attempts to get around financial obligations to home and family could lead to competitive scavenging. Bob's alleged search for money to supplement his wages and his familial obligations were seen to be unusual, but the desirability of things like scrap copper reflect the possibilities associated with acquisition estranged from the ordinary circuits that link and separate home and market. Finding scrap metal was very nearly like finding free money. Money is itself a physical material and a symbol of pure, abstract possibility with, ideally, no memory of past transactions. Yet when it comes in the form of a regular wage, the memory of money earned and the expectation of more enchain employees to social obligations, both at work and at home. Free money, unlike a wage, is ideally estranged from all memory, all interpersonal relations. According to Eddy, Bob's pursuit of free money could only be achieved by sacrificing others—in particular, the employees who spotted the scrap metal or helped retrieve it.

Money serendipitously found is not simply money, any more than scavenged commodities are mere commodities. As Mac and I walked around the slopes one windy day, picking stray paper bags that had blown away from the dumping site on top of the hill, I watched him put down his plastic bag full of paper, bend his knees, and pluck an old penny from the soil that I had barely noticed. It was scratched to the point of illegibility, but Mac carefully turned the coin over and studied it in his hands, trying to read the date or the inscriptions along the side. Tired from a day's work, we took a break from walking along the uneven ground and picking wet trash. He explained to me that he was always on the lookout for 1943 pennies. They are very rare and valuable to coin collectors, Mac said, because copper production was halted in that year to support the war effort. Copper is not just copper. The penny was not from 1943, and neither were any of the others he had found over the years at Four Corners, but he placed it in his pocket anyway and we continued working. For Mac, scavenging was not simply about avoiding work or acquiring things of value, although these certainly motivated him to do it. Rather, it was something of an end in itself.

Collecting objects in this way, paying them homage as sacred tokens of value, could be considered the quintessential form of commodity fetishism.[19] Something of this sort was clearly at work in Mac's life. Indeed, in my first attempt to understand his actions and aspirations (Reno 2009), I directly compared Mac to any mass consumer, endlessly seeking to start anew. To interpret scavenging at landfills only in terms of commodity fetishism, however, is to downplay the importance of normative market exchanges—the practice of shopping and paying for goods with money—in promulgating dissimulating representations of capitalist social relations. Finding things in the waste is not somehow less abstract (if anything, it is even further mediated and complicated by the absent presence of the waste maker), but it is different from normative commodity exchange—and so, too, are the kinds of imagining and relating it makes possible.

On a later date, I visited Mac at his new place, the one he'd acquired after leaving the house he had once shared with his mother on landfill property, now torn down by management. The coin had joined a collection of knick-knacks elaborately arranged above the television set in his new living room: old coins with the dates worn off, small figurines like the ones his mother once collected, and diamonds with slight imperfections, some of which he believed were valuable and others he merely found pleasing. Whatever else they meant to him, they were tokens of possibility. If some possessed doubt-ful exchange value as individual pieces, they still represented the significance of redeeming and reusing things that have been lost to others and hinted at what other treasures might be out there still. Even if they were not worth-while collector's items, the coins still served an ornamental purpose as a sign of generic possibility, a blending of the semiotics of waste and money. Mac was single at the time and his house was arranged neatly, in part to preserve the memory of the mother he had lost. When last we spoke, he still felt her influence in his daily life. When he first moved into his new place, he sus-pected that her spirit was trying to communicate with him through the objects in his new living room, especially the lamp he had recovered from their old house before it was torn down. This blending of mystery and mun-dane object seemed to be a source of comfort and curiosity.

For Mac, Bob, and others, the occasional material find—scrap metal, an odd coin, a possessed lamp—revealed a level of spontaneity underlying tedi-ous days and weeks performing difficult work and dwelling in a troubled home. This is also something uniquely different from the experience of shop-ping. And one does not have to be accustomed to scavenging at a landfill to be drawn in by the promise of free money. The young security guard also collected scrap metal to supplement his income and offered to take an old car engine from Timer's yard. And Timer was happy to see it go, because he looked to supplement his earnings and demonstrate his worth through other means. It is easy to see why Mac, when I knew him, would sometimes fanta-size about one day leaving Michigan and going to Arkansas to spend his days at the Crater of Diamonds State Park in Murfreesboro. At the park, he once heard on television, tourists are allowed to dig in the ground and keep any rocks they find; some people have made millions from diamonds recovered there; and, though the diamond fever that once surrounded the site has died down, the possibility of finding treasure remains. Mac insisted upon this, and for him it represented the ideal life—far from the forced choices of home and market, with nothing but potential treasure waiting in the dirt.

TINKERING

Someone unfamiliar with scavenging may depict it as a purely individualistic and utilitarian pursuit of goods, meant for people without the means to buy things new. Scavengers, according to this reading, are just like capitalist shoppers: people interested in getting a good deal on things they want. The obverse assumption, more critical of consumerism, is that an attachment to buying things new—and, by extension, an avoidance of scavenging—is associated with an inability to remake and repair things as they break down. Practices of scavenging highlight that ordinary consumption is both more active and more passive than is generally understood. It is active in that consumers are choosing to buy rather than make, and to acquire new rather than reuse. But it is also passive, because the only agency consumers exert is in their choice of what product to buy and where to purchase it, whereas scavengers are often forced to experiment with, and learn more about, what they salvage.

In addition to scavenging, I want to consider creative acts of reuse among workers at Four Corners, which I characterize as *tinkering*. Tinkering is clearly distinct from, but complementary to, scavenging. Recovering things that someone else has disposed of often means that the scavenger has the unique vision and talent to realize a thing's untapped possibilities. If successful scavenging originates from an ability to perceive value where others do not, then successful tinkering means having an ability to realize value where others cannot. Put another way, scavenging something is about a recovered item's potential for you, whereas tinkering with something is about your potential for it. In this regard, tinkering shows more clearly than scavenging a way of using objects that escapes consumption as typically imagined.

Scavenged objects may be recovered and valorized by an individual, but their involvement in human relationships is about more than their immediate utility. People relate to things, even those they literally consume, in ways that go beyond simple encompassment and destruction. At Four Corners, *re*claiming the right find could make you seem defiant, lucky, clever, or funny—provided that you successfully *pro*claimed it to others.[20]

Gendered Tinkering

There was one business in the vicinity of Four Corners that drew almost as much criticism as the landfill itself, and that was the neighborhood junkyard. Though it was fenced off and rarely used, neighbors complained about how it might devalue their property or pose various risks. Some also commented negatively on the kind of people who frequented the establishment, who were typically of lower social class.

All businesses dealing in mass waste must attend to its relative indeterminacy, as do scavengers and landfills. Junkyards earn profits by sorting through used vehicles and making their potentially reusable components available for salvage, whereas the regulations, technologies, and transactions that make up the waste disposal industry are meant to contain or lessen its potential for pollution.

I once accompanied Timer to another, bigger local junkyard. He was in search of a material component for a Malibu he was rebuilding from discarded car parts. These he had gleaned over the years from junkyards and through deals with friends and family. The junkyard also bundled commodities to ready them for sale. Cars were left in rows, organized by their make, and propped up so they could be examined for necessary parts. In this way they were made accessible to the visitor, who could remove components for purchase, as we did. In the back were the remnants of failed (and successful) recovery: vehicles that had been compacted into bricks of steel, ready for sale as scrap metal.

The junkyard was a secondhand marketplace, a place for those interested in tinkering with vehicles. The price of admission was only the skill to identify and remove what one needed. All junkyards are, thus, simultaneously places to shop and work*shops*—sites of practical expertise and know-how. Being able to tinker means being good at making things work again or making do with what's available. All acts of tinkering involve skill, and an appreciation for the open-ended potential of objects and materials.[21]

Like junkyards, landfills could be said to embody both senses of "shop"— both places to consume through scavenging and sites of expertise. Aside from the opportunity to reclaim things from the waste, the maintenance building was overflowing with remade and reused vehicles and machines. Recycling and reuse were more than company policy and propaganda. While there was clearly an explicit goal to project a "green" image—evident in the reuse of leachate, autofluff, yard waste, and methane for various ends—a tinkering ethos also permeated everyday activities and conversations:

Todd and Leon use a bra [from the garbage] to tie together two pipes they'd raised, in addition to bungee cords.

Eddy uses a plastic lid to a bucket from the garbage, wedges it under the stand that boosts the trailer higher than the hitch, preventing it from getting stuck in the mud.

"I'll send them soul chickens to Texas to feed the hungry!" [Big Daddy brags about recycling the seagulls he shoots at up top, which pester the operators]

Workers were accustomed to drawing on what was at hand to make jokes, pass the time, or fashion tools to serve a temporary need. This could serve as a concrete manifestation of competence and skill. Ambivalent as many were about scavenging, they all admired the skill that it took to tinker, and many sought to be recognized as reliable, clever, and handy.

Tinkering can occur in a variety of forms, and not all are considered equal. The meaning of tinkering has shifted over time in North America—along with, and in relation to, shopping and the two-sphere ideology of home and market. In and outside the home, gendered domains of sociability become associated with different tasks and technical aptitudes. Acts of restoration and maintenance are thus gendered and linked to the ideological divide between materially productive work and socially reproductive labor—for example, the value accorded to working on a house in comparison to house-work or housecleaning.[22]

Tanya made a habit of reusing supplies from the Roach Coach in a clever way and took great pride in concocting tomorrow's specials out of yesterday's leftovers. Timer's neighbor, an elderly woman who lived alone and looked after his family, would take apples from his yard that he thought were spoiled and transform them into delicious pies; she also converted a toilet that had cracked into a flowerpot and placed it prominently in her front yard. Another neighbor took cattails from a perimeter ditch, with permission, and created flower arrangements with them. These acts of tinkering demonstrated various skills but were not the kind that could build one's reputation among male workers at Four Corners. Among my former coworkers, tinkering was considered symbolically masculine, productive work when it involved repairing and reconstructing machines that represented aspirational middle-class values—like computers, cars, or the home itself. These acts of tinkering can be contrasted with more feminized reproductive labor like mending clothing or transforming leftovers into a new meal.

Tinkering with technologies is central to the codification and embodiment of gender in North America, as it is elsewhere. Characteristic in this regard are practices of car care, which include repairing, fixing up, maintaining, and understanding vehicles of all kinds.[23] At Four Corners, car-care activities were concentrated in the maintenance building, which was the facility's workshop and garage in many senses. A site for storing and using vehicles, tools, and parts, it also had a regulation basketball hoop that Big Daddy used, "girlie" calendars that were left lying around, and other odds and ends scavenged from up top. Gendered objects and activities were woven into the space and served to identify it as a site of masculine sociability.

In the same way that shopping became represented as a form of feminine expenditure, technical shops connote masculine work with machines and other, similar commodities. One of the most respected people at Four Corners was Roy, a senior mechanic. Roy was incredibly gifted at fixing seemingly worthless things and was regularly acknowledged for this. When the landfill provided vehicles for the demolition derby at the local summer festival, Roy would take junk cars and remake them so that they could be driven and destroyed for local amusement. Roy was also responsible for fixing landfill vehicles that would otherwise be scrapped. While I was working there, one of his most impressive feats was to create a dazzling green truck from the remnants of two pickups that had been wrecked in work-related accidents. When the project was complete, the remade vehicle looked newly purchased. For days, while it was conspicuously on display in front of the maintenance building, everyone marveled at the bright green wonder, praising Roy's natural talent.

Knowing vehicles was an accepted form of demonstrating know-how that resonated with my coworkers' conceptions of masculinity. Eddy routinely made fun of me for my inability to distinguish different makes and models of vehicles, and others would occasionally comment on the (foreign) Toyota that I usually drove. These social gaffes were implicitly and explicitly characterized as signs of my inadequacies as a man. As discussed in chapter 2, Todd's habit of damaging work vehicles was a source of regular mockery and scorn, as if it demonstrated his general inability to handle and care for vehicles—also evident in his reliance on others to fix his work truck or to lend him their truck to do his work.

Care and Masculine Pleasure

Car care was also widely recognized as a source of pleasure. As landfill supervisor, Big Daddy had the authority to do almost anything he wanted at the landfill and could take time out to choose whatever forms of supervision or work suited him on a daily basis. Most often, this was working on his and other people's trucks. While Big Daddy was not thought to be as talented as Roy, he did volunteer to fix up Eddy's mother's car as well as Todd's, free of charge, on site.

While not as respected as Roy or Big Daddy, Timer had a lifetime of experience rebuilding cars before he became a laborer at Four Corners. One of his favorite television shows was called *Overhauling,* dedicated to remaking old cars into brand-new ones. His own tinkering was inspired by work that the hosts of the show did on a 1953 Chevy, which they built out of three other cars. The rebuilt Malibu gave him something to atone for the many cars he had rebuilt and lost over the years, which he attributed to bad luck and past mistakes. It also allowed him claim personal time and space while at home, through gendered and gendering practice. In this case, the garage was primarily his domain, as well as that of his large dog Buford, and his preferred dwelling place.

Distinct from the reproductive labor of the home, garages are often sites of tool use, mechanical tinkering, and machinofacture.[24] Garages are simultaneously places where productive tinkering and risky material accumulations can be concentrated. The garage was where Timer left stray objects, either given to him or salvaged, that he could potentially make work. Timer would use working on the Malibu as a pretense for my visits, which would often conclude with drinking in lawn chairs, looking out at the landfill from his open garage. At times, renovating the Malibu helped him feel like a good father, offering occasional opportunities to teach his eldest boy how to sand down dents and do other body work. Timer discussed teaching his eldest to "touch up" cars as an important skill to pass on: "This way, if something happens, he'll know what to do." But fixing up the Malibu also gave Timer opportunities to escape from his family and spend weekend afternoons and evenings in the garage, drinking, listening to rock 'n' roll on the radio, and tinkering in peace.

At home, one of Timer's favorite things to do with the machine was "torquing it up" by revving the powerful V8 engine and burning rubber from the tires. The thick plume of smoke that filled the air and the tar-black streaks that stained the driveway were not merely signs of the engine's rotational force, tests of its performance, and conspicuous displays of its power, but also served as evocative demonstrations of Timer's skill as a mechanic. As he once proclaimed proudly while torquing, "As long as Mac [its previous owner] had

FIGURE 6. Timer's Malibu, with the landfill in the background.

it, he never smoked the tires. Now look at it! Do I know what I'm doin' or what?" Once forsaken and now partially redeemed, the Malibu objectified his own potential.

At the same time, the patchwork nature of his rebuilding effort occasionally left Timer frustrated and uncertain. On one occasion, the engine spouted flames; on another it began leaking oil profusely in his driveway. It was not always clear whether its cobbled-together parts were still good. Eventually, these continual breakdowns forced Timer to sell the car, which meant that he could not fulfill his dream of driving it to work every day to show off to others his successful masculine tinkering, as Roy could do.

Timer was unusual at Four Corners for his willingness to leave the Malibu in his driveway, where it could easily be seen (potentially by prospective buyers). Not all of his coworkers would have been comfortable exhibiting their handiwork so publicly, least of all those who aspired to middle-classness. Many would talk about going home to work on their houses over weekends and holidays, whether this meant doing actual constructive labor or merely tending to household chores and routine maintenance. Such work could not be publicly judged, except by visitors and neighbors. Unlike the marked tinkering prowess of Roy, Big Daddy, or Timer, working on a home did not make one uniquely talented, but it did provide a more significant source of self worth for many operators at Four Corners than anything they could remake through scavenging or junkyard visits.

For those at Four Corners without a mechanical background or state-of-the-art tools, remaking things was a riskier proposition and might not be taken seriously. Around when Roy was rebuilding the green truck, Eddy talked about acquiring an old car to "fix up." Few actually believed he was capable of this, and eventually he did give up on this idea. The only act of tinkering I was able to perform for anyone at the landfill occurred when I gave an old computer to Timer (one I no longer needed) and tried to get it to work for his sons to play with. Ultimately, this was a failure: the computer worked, but the boys had wanted to play a computer game with which it was incompatible. Timer had hoped they would be able to go online, not realizing the additional connections and costs involved. But he and his boys clearly appreciated the effort I made and the limited expertise that I demonstrated in knowing what could not be done. For a time, they kept and made use of the luxury commodity all the same.

For Timer, the garage—as a site of tinkering and revaluation—helped mediate between family and work, his home and the landfill. His garage was a place where boundaries could be carefully bent and redrawn between value and waste, dignity and emasculation, skill and failure. The garage seemed to make him comfortable with his handiwork, so that he did not mind failing or deferring to the expertise of others. More to the point, he gained pleasure from the anticipated social interactions that each new tinkering effort would lead to. Just as good tinkerers need an audience, they also may gain experience and joy from sharing stories of successful or unsuccessful projects.

When Timer's nephew gave him a snowblower that neither of them could fix, he immediately decided to show it to Roy. Timer regularly relied on Roy's acknowledged expertise, as did many others. Roy encouraged people to seek him out in this way. When I purchased a PlayStation 2 to play games online with my coworkers, I accidentally damaged the headset that had to be purchased to use the game. In fact, I had inadvertently sliced through a critical wire as I tried to free the commodity from the plastic bundling waste it was encased within. I was surprised when Roy, overhearing my reported mistake, invited me to show him the damage. This was the first time he had spoken to me directly, after weeks of my working there. His kind gesture felt like an affirmation of my belonging at the site and, in a small way, in his life. Such was the importance of tinkering to masculine sociability at Four Corners. And though Roy admitted that the headset was beyond repair, it was clear that tinkering, for him, was about more than being a mechanic or working for a wage. For Roy, as for Timer and many of their coworkers, tinkering was also about being a man, a father, a friend, a good person. It was about social relations and exclusions, those crucial dimensions of human life whose re-creation is at the root of any effort to make, circulate, and fix up the things we value.

MASS CONSUMPTION MADE STRANGE

Paradigmatically middle-class commodities require extra commitment—devotion to the impossible task of making them last. This is most obvious with practices like car care, which seek to squeeze out the last bit of use from vehicles until they eventually must be scrapped. But these bubbles, big and small, in which we invest our time and money eventually burst. The inevitable decline and decay of possessions may be met with anxiety and guilt as owners are forced to separate themselves from the things they have cared so assiduously for, lacking the skills necessary to make do and revive them from their fall into transience.

The inevitability of waste has bedeviled liberal notions of property and ownership from the beginning. The origin of North American–style liberalism is often traced to the writings of the English philosopher and political theorist John Locke. In the influential section of *Two Treatises of Government* (1690) titled "Of Property," Locke argued that private property relations were possible only after the problem of waste had been solved. Private ownership would be morally abhorrent, he reasoned, if some went without while others acquired things in excess that spoiled in disuse.

Bringing Locke's argument up to date, we could argue that waste subtraction mediates relations between producer, seller, and buyer but also disguises the greater social problem of unequal access to privately held material goods. If private possessions were left to rot in the open, unpreserved and uncared for, it might draw attention to the injustices of private ownership. Perhaps this is really what is so unnerving to some about junkyards, Dumpster divers, or squatters who occupy abandoned buildings. These illicit uses of someone else's uncared-for waste suggest that some own so much that their disused possessions are needlessly going to waste, while some own so little that they must make do with what others neglect. This could also account for the ways in which spoiled products are secretly whisked away to be disposed of out of sight, sometimes under armed guard, lest the mass spoilage of commodities call mass consumption into question. If widespread wastefulness is hidden from view, then the reified sphere of the market appears as if it only involves competition between companies (struggling with each other to attract consumers to their bundled goods for sale) and between consumers struggling to acquire and preserve them. Foreclosed from consideration are not only struggles between companies and consumers or between those who own and those who make things, but the fate of all those excluded from or electing to opt out of these competitions altogether, who might take a risk and retrieve food out of the mass waste stream, with or without packaging. But widespread scavenging is disallowed by the existing legal and technical infrastructure of waste management.

For Locke, solving the waste dilemma meant that choosing not to own things privately had to become the more wasteful decision. This was made possible, he reasoned, through the introduction of something that could be owned in excess without spoiling, namely money. In keeping with his ideas, many North Americans equate wastefulness with poor financial decision-making. During my research in southeastern Michigan, reconstructing and preserving one's house served as a material analogue to the widespread fantasy of 2 percent growth in financial equity. This entails a marketization of the home, which becomes a virtual object of ever appreciating value as well as a lived-in dwelling. At the time, friends and family accused my wife and me of being wasteful, in a Lockean sense, of "flushing money down the toilet" by renting an apartment instead of buying a house in Michigan when the market was so favorable to borrowers. As Locke envisioned, this excretory metaphor—of a renter wasting money rather than investing it as productive capital—conceals the other kind of wastefulness associated with private property. To ensure a return on their investment, homeowners must actively maintain and lay waste to their dwelling, thereby sustaining its material durability. But the seemingly never-ending work of repair and improvement entailed in property ownership can underwrite submission to financialized debts and assets. Aspirational fantasies of self-creation and guaranteed capital return, we now know, helped lead to the twenty-first-century housing bubble and its collapse. The simultaneous growth of landfills is a further analogue to these events, but one so removed from view that their burial of construction and demolition material, bundling waste, and obsolescent furniture and electronics goes unacknowledged, as does their continued role in maintaining the home as a commodity form. Like many banks, landfills also seem as if they are too big to fail, but our overreliance on them is rarely associated with the bubbles we create.

The desire to purchase new commodities, to achieve virtual transcendence in our material lives, has far-reaching implications. Following Locke, the cost of ordinary consumer practice—of buying things new in shops—is not only overuse of resources and destruction of Nature in the abstract, but also our forgotten social obligations to and dependence on each other. We easily forget the various, disposable peels and inessential components that commodities shed, but during their social lives they make possible a profane sphere of competition, shielded from a wider moral accounting that would consider divisions between those with the means for apparent self-creation through consumption and those without.

In scavenging, one glimpses what it might mean to resume caring for things, not as immaterial values that must always appear new so that they might be resold, but as socially meaningful products of human skill and passion, which, like us, are destined to die. This means accepting the challenge of their inevitable spoilage, as well as recognizing what people are really capable of, beyond the impossible ideal of preserving what others have made and, by implication, know more about than the consumer.

Because landfills constrain experiences of process and change, the disposability they provide also limits our sense of human potential. The reason why landfills tend to forbid scavenging is that it gets in the way of the work of quickly and efficiently disposing of waste, which is precisely what made landfills competitive with incineration and urban gleaning in the early twentieth century. Jean Vincenz, one key innovator of landfill design, was primarily interested in ensuring a productive and orderly labor process, which led him to favor large-scale, mechanized disposal over the slow and deliberate work of sorting and gleaning. At Four Corners, Big Daddy and Bob made it a rule on site not to have laborers picking paper in the vicinity of the active cells, where waste was dumped and compacted. They reasoned that this would prevent accidents, but it also had the effect of separating the least-paid workers from much of the incoming waste. Without access to the waste loads, they missed out on scavenging things of value, but they also lost opportunities to do something opposed to landfilling in principle: taking the time to recognize and learn about a particular thing.

Scavenging can make consumption seem strange. It can make the choice to buy and have things new appear as just that, a choice, and thus interrupt the tacit norms of material transfer that sustain the ideological divide between home and market. Scavenging subverts the logic of ordinary shopping by circumventing the acquisition of goods through the market and—in the case of recovered scrap metal—of money through wage labor. Mac's anticipation of serendipitous encounters with objects of (possible) value, for example, more closely approximates the earliest definitions of fetishism.[25] If, as Marx humorously noted, the bourgeoisie are the real fetishists, it is worthwhile to reconsider the comparatively enlightening potential of scavenging. Mac's encounters with uncanny objects estrange them from normal market relations. They are not outside of this fetishistic dreamscape, which clings to them still, but nor, unconstrained and unbundled, can they be evaluated apart from the particular circumstances of their sudden, unexpected appearance and recovery. Singular qualities cling to them, sensuous traces of their journey through the mass waste stream—a dent, a discoloration, a malfunction, a lingering odor—which demand closer inspection. If one manages to recover, remake, and reuse such free and tainted objects, mass consumption and wasting do not lose their importance—they are what ultimately provision these scavengers with so much waste to pick through—but the bewitching hold of these social formations lessens, however slightly, in favor of a world of equally fantastic material possibilities and skilled re-creations.

When tinkering (symbolically masculine or otherwise) is on the decline, the capacity to realize new uses out of transient materials is also lost. This, in turn, makes North Americans more dependent on disposability. If things and materials are readily disposed of, and if people cannot imagine how they might be reused and remade, then they more easily become part of mass waste streams. Yet there continue to be other, more or less troubling, kinds of scavenging and tinkering that persist throughout North America, whether or not one has access to a landfill. These include the repurposing of junk, the restoration of recovered objects and reoccupied buildings, and various other do-it-yourself projects and "life hacks." North American tinkering skills have not disappeared but are manifesting in new domains, where people escape the potentially alienating grip of mass consumption in order to relate to objects and each other in alternative ways.

Tinkering with things may offer wealth and worth outside of, or in tension with, the presumed home–market ideological divide, but it can just as easily become a supplement to their virtual, idealized separation. Fixing one's own roof might be simultaneously about doing things "on your own" and supporting others; about buying up materials for a home and investing in a source of capital; about maintaining and wasting. Many of my coworkers owned houses, or aspired to, and did not tend to see the radical social potential that I have ascribed to their activities. For them, tinkering was usually focused on the maintenance of aspirational middle-class commodities: the car, the house, and computers. Gendered stories of scavenging or illicit acquisition likewise tended to dwell on exceptional luxury goods, golf clubs, alcohol, pornography, and PlayStations—all the more worth discussing if they offered opportunities to transgress work discipline, redress perceived marital imbalances, circumvent the limits of one's wage and the constraints of the marketplace, experience pleasure, or simply imagine a better life. In this sense, scavengers and tinkerers at Four Corners and the world over are involved in the same struggles over the worth of things, dwellings, and persons as we all are. They just have the benefit of additional, extraordinary opportunities to profit from and play with stuff.

FOUR

Wasteland Historicity

MANY TOWNS AND CITIES IN NORTH AMERICA are surrounded by old dumps that were likely closed by the late twentieth century, after the introduction of new regulations. In general, human settlements leave behind evidence of occupation in the form of *middens:* material deposits of domestic rubbish and/or excrement. Archaeologists have learned to approach the waste deposits of middens, dumps, and landfills as imperfect traces of past behavior. To do so, they must take into account taphonomy, the formative process by which these traces are disposed and become available as evidence.[1] Taphonomy includes the effects of erosion, microbes, animals, and acidic soils, which can slowly transform and distort the record our remains leave behind. Because making and moving waste alters what can be known about the past, it plays an unacknowledged role in the taphonomic mediation of historical memory. Whatever else waste management accomplishes, it also divests places of their own material remains—or forces them to accumulate those of other people.

The historical remains of people from many places can be found in Harrison, Michigan, a rural township on the edge of Detroit's greater metropolitan area. Harrison hosts two landfills, one closed, one still open, both filled to bursting with the waste of other communities. Were anyone to dig up Four Corners landfill, they would have trouble knowing whether artifacts originated from Toronto, Detroit, Lansing, or from states much farther away. Contemporary landfills are usually much bigger than middens, and far fewer communities have one of their own. Waste burial has become a translocal affair, which is why many North American landfills are located along major highways.

Such unequal waste distribution is not uncommon in North America and represents the resolution to a decade-long debate about the future of waste disposal. During the 1980s, there was widespread concern that destinations for waste would be harder to find. At that time, images of a hopeless cargo ship hauling waste from New York City from port to port, searching for a place to dump, became highly publicized. It appeared that the United States was running out of dumping space. Sending waste to places like Harrison became the preferred solution.[2]

Mass waste comes to rest in places that are considered *empty*. They are empty not only because they possess raw air space to fill with translocal discards, but because they lack the means to resist the preemption of local lands for such purposes. In this chapter, I consider emptiness as an absence that is somehow represented as such, including situations where absence appears as a natural characteristic of a landscape. Places and people defined by what they lack invite something and someone to fill the vacuum. Representing landscapes as empty is therefore a highly political gesture that helps prepare them for preemption in some way. How does a place become empty? What places like Harrison lack could be summarized simply as valuable property and political influence. Explaining how Harrison acquired these characteristics and became attractive for landfills involves making use of historical evidence and narratives, but it also highlights the fraught formation processes of historical knowledge itself. Adopting an archaeological sensibility, I argue that Harrison's emptiness is related to its distinctly rural historicity.

Historicity can be understood in two ways. In the process of living, we leave material traces behind that may be preserved or lost and forgotten, including written documents, fossils, footprints, and pottery fragments. On another level, these collections may be assembled and analyzed or silenced, thereby changing how we narrate and understand the past.[3] History as lived and as told are very different but also interrelated. Not only do we act in the present with the past in mind; in doing so, we also produce the historical facts and interpretations that will guide future actions. The difficult thing about understanding places like Harrison is that few historical remains have been left behind, and even fewer have been assembled into accessible archives or collections. I argue that this paucity of historical information is not an incidental outcome of Harrison's rural status, but is part and parcel of Harrison's ongoing representation as an empty place waiting to be filled.

Waste is a memory or trace of a past relationship, and North America has always been just big enough for people to forget the past and start over. This is what rural America has meant to colonists and migrants for centuries. Emptied of their native inhabitants, rural wastelands inspire fantasies of freedom and self-creation. The frontier myth, important to the United States, Canada, and other settler societies, is simultaneously cultural fantasy and economic strategy.[4] Seen as poorly used or disused, "virgin lands" inspire development and dispossession, which allows seemingly empty places to absorb surplus capital and population, thereby averting crisis and contributing to a broader, exploitative economic dynamic known as "uneven development."[5] A place's rurality is not natural or given, in other words, but is the product of successive cycles of investment and divestiture. The fate of Harrison was sealed, in part, by the rise of industrial capitalism in the Midwest, which led to greater interregional economic integration and intensified struggles between settlements over relative prosperity and influence. A territorial division of labor resulted, whereby some places became more attractive for unpopular developments like waste sites.

As an unevenly developed rural backwater, Harrison has continually attracted the social and material leftovers of other people and places. In order to explore its production as an elsewhere for others, I focus in particular on the importance of emptiness in Harrison's history. The places where we dwell are arguably constituted as much by what did not happen and what nearly happened as by what did. A lack of ownership, occupancy, and enchantment was part of the colonization and transformation of the North American landscape, which continued alongside the perpetual displacement and decimation of Native populations. Harrison's ruralization gradually depleted competing settlements of people, businesses, investments, and even historical memory. Most of the assembled evidence of Harrison's past is located in the more populated town of Riverside, where many former residents have settled.

The site for Four Corners landfill was initially proposed in whiter, wealthier, and more populous Calvin Township, but when this was resisted, the company chose to develop the site in Harrison instead. A widely cited industry study released in California in 1984 found that resistance to disposal facilities is least effective, and waste sites are more easily sited, in poor, older, and less educated areas.[6] But even companies that do not seek out peripheral areas are likely to locate waste sites within them by following the path of least resistance. The landfill company may not have been aware that Harrison was poorer, more African-American, and less populated, for example, but it was near the highway, land was cheap, and the farmers gave in. However, as I will show, these characteristics are historically interrelated.

Out-of-the-way places like Harrison can be acquired cheaply because of their rurality. Among the support provided by Harrison's landfills to the surrounding community have been basic infrastructural services otherwise denied to its residents, including a fire department and an expanded sewer system. Harrison's rurality has continued to attract the schemes and fantasies of a variety of actors and institutions who seek to take advantage of the seemingly natural setting, whether as a repository for foreign garbage, rare wildlife, dead animals, and freed pets or as a serene place to escape from other regions and locales. In some cases, the township's landscape is preempted in ways that present challenges to people who live or have lived there. When changes to the land are extreme and relatively permanent, as with landfills, residents may struggle to make sense of what once familiar places have become and accept the strange new possibilities they introduce.

The ambiguous relationship between the two types of historicity complicate attempts to understand the past in the best of circumstances but are especially fraught in the context of places like Harrison, where things are sent to be forgotten and people go for a fresh start, unencumbered by history. The uncertain gap between making history and documenting it is an uncomfortable one to occupy. In the process of researching Harrison, I also became complicit in the contested historicity of the township's past, and altered local assemblages and narratives in the process. Beset by these difficulties, what follows does not obey the conventional rules of linear historical narrative. I present details of the past out of sequence—which is how both I and my informants actually learned and experienced them—and counterpose official histories of the region to local narratives and those local narratives to each other. I intend for the resulting account to approximate the actual open-endedness of time and place in Harrison, and to highlight the many possibilities for telling the past that remain or have been lost.

REPRESENTING RURALITY

By the early twentieth century, Harrison's perceived emptiness began to attract migrants. Becoming rural is not simply about what a place lacks—whether infrastructure, businesses, or population—but also about what is made possible as a consequence of such uneven development. As part of its emergent rural position vis-à-vis neighboring locales, Harrison acquired a bucolic appeal that has attracted people from the Detroit area, the American South, and Eastern Europe over the years. To many of them, the township would have seemed secluded, which meant having land to grow food, experience nature, or be left alone. Various newcomers defined Harrison throughout the twentieth century, providing opportunities for local revitalization, creating many of its most enduring civic organizations and businesses, and leaving lasting impressions on rural life in the township.

During the 1920s and '30s, Americans increasingly left metropolitan areas for smaller communities along the urban periphery. The movement of elite urban migrants led to a process of suburbanization, which had already begun in parts of industrialized Europe decades earlier. This was encouraged by the decentralization of industrial production associated with the booming automotive industry centered in Detroit. After 1925, most industrial development in the Midwest took place on the fringes of urban centers like Chicago and Detroit, giving what had previously been smaller towns an opportunity to expand (Kenyon 2004: 20). Riverside, the more populous village to the north of Harrison, was well placed to attract elite migrants from the city because it was situated on a main road, near three large automotive plants built in the thirties and forties. It grew by almost 115 percent between 1929 and 1949, as people relocated to live within the new industrial center of the country.

Also during the interwar years, tens of thousands of Upland Southerners moved north to find work in the auto industry, some of them directly recruited into factories to make up for frequent labor shortages. Harrison was not the most popular destination for southern migrants, but it did serve to absorb surplus population, offering relatively cheap housing for the surrounding area, including three trailer parks by the sixties. Some southerners were attracted to the township because of its open farmland, similar to what they had known in the South. These newcomers played a seminal role in redefining Harrison's brand of ruralness that continues to the present day. The term *hillbilly* originated in the northern states to refer to migrants from the Upland South during this period. Representing a racially marked form of whiteness, hillbillies challenged the racial and class fabric of the northern and midwestern states.[7] While Harrison did not attract very many elite or hillbilly migrants, its population did grow. And like neighboring locales, Harrison gradually became "post-agrarian" as its social fabric was shaped less and less by agricultural production and farm life (Salamon 2007). At first, factory work was something done to contribute to the household income and supplement farming. But between 1930 and 1950, the number of farmers gradually declined, despite a booming national farm economy in the postwar years, until more than half of the population was dependent on the automotive industry for income.

Today, Harrison is known for its hillbilly influences. Residents are routinely mocked for their hillbilly qualities by outsiders in neighboring towns. Workers at Four Corners labeled locals hillbillies and made jokes at their expense, including allegations of incest, stupidity, southern-sounding dialects, and poverty. On occasion, Four Corners would offer free dumping for locals, which was jokingly described by landfill employees as "hillbilly dump day." Even the name of the township was occasionally pronounced "Hairsun," with the accent intentionally emphasizing the first syllable, as a way of indexing and mocking local speech. Among the transgressive activities historically associated with Harrison's southern hillbillies are cockfighting and moonshine stills. Some claim that Harrison served as a site of illegal operations for the notorious Purple Gang that ruled the crime world of Prohibition-era Detroit. Residents familiar with these stories are prone to say that illegal pastimes were more prominent social activities during the Great Depression. At the same time, Harrison residents of different ages claim to have knowledge about where stills can be found today, have personally attended a cockfight, or can, at the very least, point out homes where cocks are raised for regional competition. Though the idea of the hillbilly remains a part of local discourse, it is now used more generally, and not to mark and segregate problematic whites as it once was. A far more important aspect of social identity in contemporary Harrison concerns whether one is Polish or Black. This also serves as a social index of approximate arrival of one's kin, whether this occurred before or after the hillbillies.

If Upland Southerners came to the Detroit area searching for employment in the booming auto industry, others had been attracted to Harrison more for its ruralness, specifically the availability of cheap land and relative seclusion. As early as 1885, Polish farmers who were having trouble acquiring property in neighboring townships found Harrison more open to their presence. By 1930, almost a quarter of Harrison's population was foreign born, twice the average percentage in surrounding townships. When a Catholic church came to Riverside to serve Harrison's Polish populace, it was initially met with a great deal of local resistance. Rumor has it that the church achieved local acceptance only through a clever land deal. More so than the Upland Southerners who would come later, Eastern European farmers became influential in Harrison's social life and, as the income of farmers rose in the postwar years, supplied important new businesses and civic organizations to the area. Two of the more significant for Harrison, throughout the twentieth century, were the Polish Community Center (PCC) and the

Harrison Skating Rink. The former served as an important site for the local chapter of the Polish farmers' association, giving them a place to congregate and plan community events, but it also provided a dance hall and reception center, where many local weddings were held in the postwar years. In terms of its importance as a site for local gathering, the PCC was matched by the Harrison Skating Rink, owned by two generations of a prominent Polish family, where many young people associated and met their future spouses from the forties to the seventies.

In recent years, the loss of the PCC and the skating rink led the current political leadership of the township to seek alternative ways of bringing the community together and, in particular, of drawing younger families with children into the area to raise property values and increase the tax base. To that end, both Harrison and Riverside have developed separate proposals for a community recreation center that would provide a gym and activities for young people of different ages as well as young, middle-class adults. Interestingly, such a place already exists at the border of Harrison and Riverside, as part of a new Evangelical church. However, it has largely gone unnoticed by politicians in both communities, which is almost certainly related to a tendency for political leaders to neglect the contributions of African-American churches to the community in general since the fifties.

While the PCC and skating rink are now recalled fondly by many residents for their contribution to bringing together Harrison as a community, both at one time served as important sites of racial segregation and struggle. The African-Americans who arrived in Harrison in the postwar years were seeking, much as their Eastern European and hillbilly predecessors had, a truly rural setting where they could establish themselves and live independently. These stories are less well known in Harrison, generally, and have to be sought out from living African-American residents. For example, Marie Willis was twenty-one when she and her husband bought a house in Harrison in 1953. They had heard about the township through their congregation in Detroit. Their minister had brokered a deal with a retiring Polish farmer for eighty acres in the northeastern part of the township, near where the village of Beebe had once been. It was the preacher's vision that the whole Detroit congregation move onto this land, where a new church was being built, in order to establish themselves in a better environment, outside the city. Not all followed his lead, but six other families moved to that same block, each with five acres.

Many of the church members, Marie recalled, had originally moved to Detroit from the South in the thirties and forties. Her own family had moved their grocery store north from Georgia when she was a child. Importantly, all placed a high value on being able to produce their own food, as a consequence of having lived through the Great Depression. Though few had ever been farmers or, Marie admits, had the faintest idea of how to tend a garden, the idea of settling in the country and being able to produce their own food held great appeal for those who moved to Harrison from the city. Although they were not immediately welcomed and held the status of "newcomers" for years after they arrived, Marie says the different families in her neighborhood received a great deal of assistance from one another, as well as from some of their white neighbors with more experience farming. She remembers when the general assembly of the church would meet for ten days in June, an event that culminated in a picnic called "the Feast in the Wilderness" where food was gathered at the Harrison church and publicly distributed, something still practiced to this day.

Following this initial wave, an additional thousand African-Americans migrated to Harrison from Detroit between 1940 and 1960, eventually constituting nearly 20 percent of the township's population (compared with less than 9 percent in neighboring areas). Abel Watson was not a member of Marie's church but came to Harrison with family in 1956, also with an interest in having his own garden. Like many other black southerners, he originally moved to Detroit for the promise of employment and greater racial equality, but he had a strong desire to settle in the country. Having lived through the Depression raising cotton and producing syrup in Alabama, he became set on this goal as a young man: "I'm gonna get a place out in the country. When I'm too old I'll raise a garden and live on that." After being drafted into the Army during World War II and narrowly escaping combat, he returned to Detroit as a self-described "hustler" and managed to get enough odd jobs and factory work to build his family several houses on a Harrison plot. He eventually became known as a local school-bus driver for one of Harrison's schools, as well as the president of the NAACP for the area, though now he's retired from both, "livin' on the master's care" and the help of his children.

For Marie, Abel, and other black "newcomers," Harrison's ruralness meant abundant land and a peaceful, tight-knit community. At that time, it was very difficult, if not impossible, for African-Americans to acquire property in the village of Riverside. "I'll tell you 'bout Riverside," Abel said to me. "Riverside was a place if you ever know what discrimination was, it was full of it." He recalled when a racially mixed couple managed to acquire a home in Riverside in the mid-sixties and were quickly and quietly pushed out of town. Some of the Polish farmers in Harrison were more willing to sell African-Americans property. However, for decades, the southern half of the township was unavailable to nonwhites. Black residents called the road that racially segregated the township the "Mason-Dixon Line," and it remained so until the late sixties. Since at least that time, there have been rumors that the Ku Klux Klan has members living in Harrison. While I found no evidence to substantiate this claim, over the years there have been incidents that would qualify as hate crimes, such as the appearance of lynched effigies left hanging in public parks. It is probably no accident that these effigies were found in the mid- to late sixties, around the time that African-Americans began to protest the segregation of the Harrison Skating Rink. For decades, the rink was open to black youth only on certain nights of the week, allegedly because they had "different taste" in music and dance. Similarly, African-American weddings and events had long been excluded from the PCC.

Before local activists would successfully reverse such exclusion, Marie and others from her congregation had begun the Harrison Civic Society (HCS) in the late fifties. The HCS was established to promote racial equality in the area and provide black residents with a place of their own, equivalent to the PCC, where they could gather. After several decades of influence in the town, members of the HCS became very prominent in Harrison's civic life. One member, who became the first full-time policeman and helped build a more professional police department in Harrison, is generally regarded as one of the township's most beloved civil servants. The HCS building itself still stands, and, through Marie's continued leadership, became a site of regular community activities. In 1967, the HCS building was the site of the first free health clinic for disadvantaged families, made possible with the aid of a federal grant. Though the clinic is now gone, the HCS building still serves as a regular meeting place for different social groups and annual events. In February 2006, I joined two dozen or so local residents at the HCS, white and black, to play bid whist and watch the Super Bowl taking place nearby in Detroit. While it is no longer what it once was, the HCS still brings people together and Marie is constantly dreaming up new ways of interjecting more youth and more money into the aging organization.

Like Upland Southerners and Eastern Europeans before them, African-American migrants to Harrison became part of its unique rural fabric. By the eighties, African-American residents and their new neighbors had broken down many local racial barriers and transformed the township's image to one of a rural haven of community and safety, radically opposed to the police brutality, riots, gentrification, poverty, and crime that dominated local and national images of Detroit in the latter half of the twentieth century. By this time, Detroit had become increasingly empty as well, undergoing a dramatic decline in population. While this emptiness has devastated Detroit to the present day, it is nonetheless associated with a clearly defined, hegemonic narrative of growth and decline. Harrison's consistently low population and stunted growth, by contrast, can be, and is, narrativized in multiple and sometimes contradictory ways by outsiders, newcomers, and locals.

CHRONOTOPES OF EMPTINESS

For some of the people I spoke with on the subject, the history of Harrison is a person: Phyllis Kettle, an octogenarian. Phyllis's family had been in Harrison for generations, prior to the arrival of newcomers from the city, the South, and abroad. A retired teacher and a resident of Harrison for most her life, Phyllis wrote the only existing book on Harrison's local history, researched and written as part of the celebration of the national bicentennial in 1976. Due to the circumstances of its production, Phyllis's account commemorates local history within a nationalist frame—devoting considerable attention to the many war veterans who had served in the community, for example. In this way, local history becomes an expression or realization of history writ large, of events on a national and global scale. By Phyllis's own admission, there was much local history that had to be excluded from her official account. For one thing, lore and rumor had to be left out in order to achieve the dispassionate narration and objective fact-assembly associated with history writing as a genre. But even established facts can prove impossible to narrate, or too personal to share.

All of this became clear when Phyllis and her husband Tony offered to take me on a tour of Harrison in their car. They wanted to show me historically meaningful places, though most of the sites we visited were remarkable for their *lack of traces* of the past. During our tour, Phyllis proved adept at recalling local history, despite all that was missing from view. This gave a sense of moving backward and outward through time as we drove, beyond what was (or was not) immediately perceivable. Past events were recalled from movement across the landscape itself in a way that could not have been duplicated in Phyllis's history book. Had she written a novel, Phyllis might have found ways to chronotopically fuse time and space and perhaps better approximate the moral and aesthetic dimensions of encounters with the past.[8] In her case, the when-and-wheres interwoven into Harrison's landscape provoked familiar historicist sensibilities, like nostalgia and regret, but also a feeling of obligation toward the dead. At the start of our tour, we visited an old graveyard where recent restorations were being sabotaged by badgers. Despite this unfortunate sight, Phyllis was able to identify the graves of people from prominent families in the town. Perhaps struck by the badgers' nasty work, she took the opportunity to remind me (and not for the last time) that many people in Harrison were related to one another, that everyone was someone's cousin, and that I should therefore tread carefully when digging up the past.

The Struggle of Localities

The next stops on my tour with the Kettles were two country-road intersections where the villages of Ellisville and Beebe had thrived from the antebellum years through the Great Depression.

When Harrison was first established as a municipal township, the newly established state of Michigan was not yet integrated into a national, capitalist economy, but was predominantly engaged in agricultural production for limited regional exchange and local subsistence. Harrison was founded in the midst of the devastating economic depression that accompanied the Panic of 1837. The growth of its population at this time may have been aided by the plummeting price of land in the West, which would have made farmland more affordable than in the past. Places like Harrison could survive the depression in large part because regional exchange provided a source of revenue for small farmers, detached from a reliance on eastern commerce. Ellisville and Beebe were already established by this time, situated along the territorial road leading to Detroit. Neither village was particularly large, but they contained the majority of the township's population. By the time of Reconstruction, both possessed post offices, general stores, mills, churches, and schools.

All of that is now gone. The last thing to disappear from Beebe was a family-owned grocery that was gone by the early thirties. Ellisville maintained a school, which Phyllis had attended, before the districts became consolidated and Harrison's students were bused out of town. Many of the structures that once stood are now gone; a few houses have been built at the intersections where the villages had been, but nothing is left that would indicate that the empty fields and heavily wooded plots were once active centers of social life. There is still evidence of the foundations of some of the vanished buildings, but knowledge of where they once stood is preserved only in faded photographs and the memories of a select few residents like Phyllis.

This past was lost, piece by piece, as part of Harrison's transition from one small collection of settlements among many into the rural periphery of more prosperous centers. Over time, the buildings were moved or burned down; the businesses closed and were supplanted by others closer to new population centers. One mill was donated (or sold) by the town to Henry Ford's Greenfield Village, where it is included as part of an authentic frontier aesthetic marketed to tourists. Some feel that Harrison residents have lost an important sense of themselves, partly as a consequence of this active forgetting of the recent past. In Riverside, by contrast, a historical society meets regularly to discuss local history, an annual tour takes interested residents around to local houses of historical interest, and important sites are preserved to maintain a connection with their past.

Harrison's transformation into a rural locale (and the forgetting and naturalization of this transformation) were not foregone conclusions. Those who talk to older residents, or look through the archives in Riverside's local museum, learn that it could have been otherwise. Long-term residents of Harrison often relate the decline of Beebe and Ellisville to the loss of a proposed St. Louis–Detroit railroad depot in the 1880s. This would have connected the center of Harrison to one of the largest routes of interregional trade in the Midwest.[9] Railroads changed the shape of national markets by bolstering demand; with geographic distance less of a barrier to commerce, exchange value was able to circulate relatively unmoored from the material limits of commodities. Michigan railroads connected with lines running west to Chicago and northeast to the coastal cities, which made them a leading supplier of grain, copper, and hardwood timber by the 1870s. Improved transportation infrastructure placed more emphasis on the political–economic relations between different locales, whether neighboring or distantly connected, providing a more systematic patterning to uneven development.[10]

The new train depot was originally intended for Ellisville, but the railroad line deviated from its originally planned route and the depot was placed in Riverside instead. Harrison's more recent residents are unlikely to know this story. Even if they are aware, many now tend to think of themselves as part of Riverside (where they grew up, go to school and work, do their shopping, and attend church) and are typically less inclined to describe the relationship between the two locales as one of competition. This is true even of African-American residents, who remember a time not so long ago when they were unable to buy property in Riverside. In my experience, those who remember tend to be very reserved about making negative comments about their neighbors and friends in Riverside, as are people in Harrison in general. Nevertheless, in a solitary paragraph written by Phyllis to conclude a chapter in her short history of Harrison, she writes, "Gradually over the years, both Ellisville and Beebe began to disappear; Riverside becoming the shopping center due to its having a railroad." In statements like this, an implicit connection between Harrison's loss and Riverside's gain is evoked. Phyllis is right to make this connection, as are other residents of Harrison who know the story. Their sense of having inherited a town that has *lost* reflects the legacy of the rise of nineteenth-century American capitalism.

This accounts for Riverside's historical success in competing with neighboring locales for resources, population, capital investment, and general recognition. With the emergence of American industrial capitalism in the post–Civil War era, the success of one center became directly related to the failure of another, a sign of its own lack of capital investment. Such place-level competition would have been readily apparent at the time. In 1886, the first newspaper of Riverside identified competing with neighboring communities as a collective goal: "Riverside bids fair to hold her own in the struggle of localities to draw trade, and become a social, educational, and literary center." The editorial went on to connect this strive for local dominance with the creation of a newspaper: "Nothing will so much aid a community in this struggle as a paper devoted to the interests of the people." The editor was right to be confident. Unlike Harrison, Riverside was considered prime farmland during the height of midwestern migration and, even more important, was strategically located on a major river.

As late as 1883, a detailed map of Harrison showed the railroad depot in Ellisville, so confident were some that Harrison had a prosperous future. Instead, from 1890 to 1920, the population of Harrison declined by 17 percent while neighboring Jackson Township and Riverside continued to grow by that same margin. The uneven development that came after the Civil War in southern Michigan was fairly continuous with that of the antebellum period, and so it was with Harrison. Riverside's short-lived stint as a transportation hub was eventually overshadowed by the spread of Ford automotive factories into the hinterlands of the Detroit area, particularly alongside major waterways and routes of travel, which further shaped the regional division of labor between locales.

Without Ellisville and Beebe, the inhabitants of Harrison are largely dependent on wealthier and more populous centers for news, religious worship, consumption, education, civic membership, and employment to the present day. Riverside's victory in this "struggle of localities" is complete. Even the Kettles eventually relocated to Riverside to retire, bringing Phyllis's knowledge of local history with them. Many current residents are unable to distinguish their home community from that of Riverside, and some newly arrived youths are unaware that a distinction exists. They tell others they are from Riverside, since that is their zip code.

"Abode of Wild Beasts and Wild Men"

Next, Phyllis and Tony drove me to a privately owned patch of woods containing remnants of the old Territorial Road, which once cut across the township and heads on toward Detroit. In some ways, the differential placement of railroads shadowed the growth of roads within the Lower Peninsula, which meant privileging larger centers and routes of regular intra-regional traffic over others. Neither Ellisville nor Beebe was located on a major road (in fact, it was not until the 1970s that the township had close access to a major highway, which coincided with its largest growth in population and the growth of its landfills), whereas Riverside was located on the main thoroughfare leading to Detroit. The Old Territorial Road was known as the "Indian trail," an uncommon reference to the region's former inhabitants and, possibly, a literal remnant of their transformation of the area prior to white colonization of the Old Northwest.

Harrison owes its current shape—both the dimensions of its borders and the density of its population—to struggles over the Northwest Territory during the early years of the American Republic that involved displaced indigenous groups, European empires, backcountry settlers of diverse origins, ex-soldiers, and elites from the East. The rise of the American Republic brought with it a new way of preempting lands and inhabitants to the west. New settlers believed they were claiming land that had gone to waste through indigenous and French misuse. In European and indigenous hands, the Old Northwest had become a "waste wilderness," as one pioneer put it, "the abode of wild beasts and wild men" (Barillas 1989: 7). This was also the perception of American elites back east, though they tended to view the new settlers in a similar way. Otherwise known as "white savages," backcountry settlers tended to ignore official property treaties with Native American confederacies and the British and occasionally defied Philadelphia through open revolt.[11]

The Northwest Ordinance of 1787 provided western territories with a path to statehood by systematically dividing up ceded or seized Indian lands for sale and resettlement. Some characterize this process as a form of internal colonization, whereby native and nonnative alike were subjected to the colonial designs of New York and Boston elites, though to varying degrees.[12] The ordinance readied the area for propertization and exploitation at a time when the early Republic was at its most vulnerable to dissension and new lands for settlement were highly valued. And though the "wild men" of the Northwest were often placed in opposition along racial lines, they also colluded against government decrees. Some Ottawa and Potawatomi managed to remain on their ancestral lands by adopting agricultural practices that integrated them into the regional frontier economy. By the time agents of Michigan's Superintendency and the Jackson administration pushed for removal in the speculative rush of the late 1830s, local settlers had become reliant on the annuity payments that members of the Ottawa tribe received from the federal government for land cessions. Their close ties with Native American inhabitants played an important role in indigenous resistance to federal policy (McClurken 1986).

In order to rule the division and distribution of a disputed territory from a distance, the American government and powerful business interests had to make the West knowable. Prominent architects of western expansion, like Thomas Jefferson, hoped to control the colonization of the Northwest in such a way that all "savages" would become citizens. The practice of farming private plots of land was meant to give the wild men of the Northwest the character to govern themselves as their territories became self-governing states.[13] Newly colonized areas were organized into townships before they could be divided between competing settlements. Harrison was given its current shape following the terms of the Northwest Ordinance, whereas neighboring Calvin and Jackson were gradually subdivided by growing towns. Unlike the more-or-less organic growth of the original colonies, colonization in nineteenth-century America was deliberately shaped according to enlightenment categories of spatial and political order.

The now familiar borders of townships and farm plots erased older forms of movement and dwelling from the landscape. Today, Harrison's Indian trail/Old Territorial Road is barely visible, except for those who remember a time when it made sense to travel across town on foot to go to where the school and the store once stood. Phyllis and other former residents recall regularly using the trail as a shortcut while growing up. But as tractors and foot travel gave way to automobiles in the early twentieth century, the grid-like borders of the Northwest Ordinance became further concretized as roadways. The shape of movement through Harrison has been further reinforced through the more recent installation of phone, electric, and sewer lines. Our tour through the town's past was literally constrained by the paths of its colonial preemption.

County Sand and Gravel

One of the last stops on my tour with the Kettles was the old County Sand and Gravel Landfill, now covered with dirt and grass yet, underneath the calm surface, steadily pumping biogases into a nearby power plant. This was the only business we stopped to see, though we did so largely for my sake.

County Sand and Gravel (CSG) began operation in Harrison in 1964, not far from where Ellisville once stood, in the southern half of the township. It was initially just a local dump and got its name from the preferred method of waste disposal at the site, which was to dig a pit, dump garbage, and burn and bury it. The dump was locally unpopular and was eventually shut down by the township supervisor in the early seventies. In 1975, the wealthy and well-connected owner of a Detroit hauling company, Michael Bruno, purchased the dump and converted it into a state-of-the-art sanitary landfill. Along with his hauling operation, it was one of the cornerstone businesses of County Services, Bruno's private company that specialized in waste management until the early nineties. Prior to that time, there were a greater number of landfills in the United States, which meant that there was a smaller market share for each waste site. The landfills that were sited in the fifties and sixties reflected this, and the companies that owned them frequently adjusted to the competitive marketplace by acquiring other landfills and hauling operations as well.

For reasons both legal and illegal, some firms did better than others. Because of the structure of the waste market and the influence of industry insiders on the determination of prices for hauling and dumping, the American waste industry became susceptible to corruption and a popular front for organized crime in various parts of the country during the twentieth century. The association between waste management and the mafia remains, due in no small part to television shows like *The Sopranos*. When I told friends in Michigan that I was going to "work in waste management," some joked that this was a euphemism for joining the Mob. To this day, Michael Bruno is believed by some to have an affiliation with Detroit's once prominent mafia families, an allegation that some use to explain the mercurial rise of his company despite a competitive marketplace and repeated environmental infractions. County Services began as a small, Detroit-based hauling company and, over the course of four decades, went on to acquire eleven landfills and several smaller hauling companies, an operation worth $750 million as of 1998, when it was finally sold to a bigger waste corporation.

At the time of CSG's closure, it covered 240 acres and contained a reported 14.25 million gallons of waste, nearly a third of which was buried in its final four years. Acquiring all that waste during its forty years of operation made CSG a nuisance to many of Harrison's residents, though it clearly affected some more than others. Because it was sited in the middle of the township, the traffic heading to the site clogged the main road through town, spreading odor and noise in the process. Although this was much worse in the last years of its operation, there is evidence of organized local opposition to the landfill at least as far back as 1975, when Bruno reopened the site.

Even after the landfill closed, CSG continued to create problems for its immediate neighbors. Its original design was not meant to guard against the leaching of landfill liquids off site. In an attempt to correct this problem and fulfill the terms of its postclosure agreement, County Services paid for the construction of a slurry wall along a corner of the oldest part of the site, which was susceptible to leakages through surface water runoff. This proposed solution led to a lawsuit in the late nineties, when the now closed landfill was accused of inadvertently flooding the basements of several neighbors. Ultimately, the landfill paid to install sumps in the basements of nearby residents and the suit was settled out of court.

There are other ways in which Bruno and his company's preemption of a portion of Harrison was ambivalently connected to more "neighborly" activities of community betterment. Despite its problems, some residents, including Phyllis, claim that County Services was always very good to its neighbors and compare it favorably to the newer landfill, Four Corners. CSG did play a significant role in the community for a time, employing residents and sponsoring local activities and events. For years, Bruno was the largest sponsor of Harrison's annual Summer Fest, a parade and carnival that is enormously popular in the area. CSG's sponsorship of the festival was passed on to management at Four Corners later on, who continue to provide much of the budget for the event as well as supplies and other services.

Too Much Change and Too Little

For Phyllis, Harrison's present landscape is chronotopically linked with pasts that are largely gone. She pointed out empty lots that had once contained prosperous businesses, and houses that were formerly inhabited by eccentric families. Here was the town's only mansion, where the children once had to sleep in the basement to avoid damaging the fancy furniture. Here was the once notorious house where spiritualists had regularly gathered, people said, to conduct séances for the dead. Here a local mortician running the new town graveyard was caught replacing expensive caskets with cheap, flimsy boxes. He was found out when animals began digging up human bones that had floated to the surface during a particularly wet season. Phyllis showed me where. Some of these details she hesitated before repeating, reminding me more than once that many people in the township were related and that a story about one could offend many others. She let me accompany her to these storied places, but the way she spoke about them, as well as her refusal to include them in her book, reflected ethical commitments to the living and the dead who still haunt the landscape.

There were other sites, more recent wheres and whens, about which Phyllis had far less to say. She remained uncharacteristically silent, for example, when we passed by Harrison's dozens of small churches, most no more than a few decades or years old. Phyllis attended church in Riverside, where she lived, along with many other current and former Harrison residents. Most of the churches in Harrison are attended predominantly by African-Americans from the greater Detroit area, a notable disparity in a township where 80 percent of residents identify as white. As she spoke, Phyllis mused about "slower times," when people in Harrison knew their neighbors and everyone helped and trusted one another. She contrasted this image of Harrison's past with its lower-income trailer parks, now several decades old, where crime is more common because "people live too close together." By contrast, other, younger residents described the trailer parks of the seventies as a close community or neighborhood. My coworker at the landfill, Mac, was a former resident of one of them. He later took me on an alternative tour, not unlike Phyllis's, where he bemoaned the recent emptying of people from the trailer parks and connected this to their fall into disrepute.

Roaming and observing one's surroundings is quite ordinary, but when these are rich with chronotopic fusions of space and time, the result may prove traumatic. Phyllis's memory momentarily failed her when we arrived at the corner where her family's farm once stood. Her deceased parents had owned eight acres of fertile land on the southwest part of town, down the road from the intersection that formerly belonged to Ellisville. The remaining traces of Ellisville have vanished, new houses have sprung up along the road, and very few of the residents possess familiar faces or names. As we passed by, Phyllis hesitated when trying to recall certain details. She stuttered, trying to point out where her uncle's house had once stood or recall when certain events had occurred. This was so uncharacteristic, given how easily she had discussed many aspects of official and unofficial history that day, that Phyllis felt compelled to offer an explanation for her lapse in memory. It had been too long, she said, and changes to the place itself, her old family farm, were partly responsible for her absentmindedness. "You see, we don't come here because it doesn't mean anything anymore," Phyllis said. "Everything has changed," Tony added. "Too much change," she agreed.

Despite her initial disorientation, Phyllis did eventually begin to tell us family stories. She talked about the big fire that had nearly cost them their house; about huddling inside while tornados passed by on the fourth of July; about the Indian artifacts they would find while plowing the field; and about the swamp where they used to raft in the summer and ice skate in the winter. "We lived in the swamp!" she exclaimed.

The history and historiography of Michigan actually owe a great deal to swamps, which is associated with a widespread claim that the state changed too little, remained empty too long. After the Revolutionary War, early travelers described Michigan territory (which included Wisconsin and parts of Ohio and Minnesota) as an impassable and disease-ridden swamp. Wetland could be drained, and much of it was in the Detroit area, but to Americans and Europeans of the early nineteenth century, swamps and the infectious "night air" they produced were more than a barrier to cultivation—the atmosphere they generated was believed to contain disease-causing miasmas. In the early nineteenth century, there was also still widespread belief in the process of spontaneous generation; conditions of decay, it was believed, begat life forms that pestered and plagued humanity, such as mice and fleas. Living near old wasteland was, among other things, a potential source of further pollution and pestilence. This perception was so widespread that millions of acres of Michigan Territory were removed from the federal bounty-land act, which was meant for soldiers of the Revolution. Another important reason often given for the supposed delay in Michigan's colonization was the treacherous Black Swamp outside present-day Toledo, which prevented overland access to the Lower Peninsula by backcountry settlers. Until the opening of the Erie Canal provided an alternative route in 1825, Michigan grew slowly, while colonists flocked to Ohio, Indiana, and Illinois in record numbers.

Michigan's population increased by 537 percent in the 1830s, partly because of descriptive maps that suggested the area was promising for aspiring farmers. The population of Harrison also grew during roughly the same period, from a few dozen to a few hundred in two decades. But at its founding (late by the standards of its county), the township still had a population of less than 228 and, at the turn of the century, less than 1,500. Until the 1960s, when newcomers were attracted to the township for its rural appeal, Harrison consistently held the smallest populations by a significant margin in an otherwise populous county. In prominent maps of the 1860s and '70s, much or all of Harrison was depicted as submerged in one large wet prairie, the only one of its size in the whole of southeastern Michigan. In an 1871 booklet for "settlers and land dealers," almost one-third of Harrison is labeled "marsh," including both of the areas now occupied by landfills. Whatever the former resemblance between map and territory, these depictions were at odds with actual conditions on the ground: farming families had adapted much of the area labeled "marshland" for cultivation by the 1870s, including the Kettles. It would seem that eastern maps did not change enough to represent the changes in Harrison.

Farmers like the Kettles did not drain all their swamps but lived with them, as Phyllis described. But negative representations of swamp as wasteland had already indelibly shaped the township's future. Harrison was one of the last places in the area both to begin commercial farming and to stop relying on it as the principal means of employment. It remained predominantly an agricultural community until the mid-twentieth century, when manufacturing work ushered in an uncertain post-agrarian phase in which it is still mired. By that time, competition with neighboring locales and a changing national economy had furthered its ruralization. Looking at Harrison's landscape today, American colonization of the Old Northwest and Michigan statehood seem to have been retroactively guaranteed, as does Harrison's rurality in relation to neighboring communities. Only unofficial memories provide a glimpse of a world that could have been otherwise.

Traces of the past are now largely gone from Harrison. The Kettles' old swamp has been filled in, just as strangers now live in the house that had been part of Phyllis's family for two generations. And she couldn't help but comment on their poor stewardship: "They closed in the porch; that's not the thing to do!" For Phyllis, it is these recent changes that challenge memory and meaning, more so than the disappearance of official town history, to which she has devoted many years of her life. While it distresses her to see the old graveyard uncared for and the old Indian trail blocked from public access, it is the minor alterations of recent decades that keep her from traveling deep into Harrison more often, into a recent and far more intimate past.

RURAL TAPHONOMY

Anyone who visits a familiar rural area in the United States is impressed anew by a paradox: the countryside is becoming emptier, and the countryside is filling up.

JOHN FRASER HART (1975: 194)

In Harrison, there were many sights to elicit that strange combination of historical memory and amnesia evident in Phyllis's reaction to her childhood home: there were the grass-covered lots that once contained popular bars, closed down in the seventies; the rebuilt general store boarded up by old Matt Robinson's widow and, rumor has it, still stocked full of groceries; the large, eighty-acre lot purchased by Marie Willis's nondenominational church in the fifties, now fenced off from public access; the 900-acre wetland preserve created in the mid-nineties on land drained for cultivation by generations of farmers; or the once segregated Harrison Skating Rink at the center of town, adorned with a "For Sale" sign. In each of these cases, the familiar gradually fades away or is wiped from the landscape, while the decay of Harrison's diverse pasts is tied to the emergence of some unknown future.

Increasingly, Harrison's landscape appears, to its former and current residents, as if preempted by outside forces, at a remove from their collective and personal goals. This is no less true in neighboring rural communities. In general, the rural Midwest has become more subject to outside pressures since the national farm crisis of the eighties. The smaller the locale, the more likely it has become beholden to the demands and fantasies of people from larger settlements nearby.[14] In the case of Harrison and similar rural areas, cheap land, relatively weak public resistance, and a financially strapped town government are all attractive to the least popular of development projects.

In the seventies, the state of Michigan began construction on a new highway intended to address the higher traffic volume flowing between the more populous suburbs of the greater Detroit area. Many people in the area recognize the coming of the highway as ushering in a new phase of developments in the area, including more speculative business schemes. The highway did not pass directly through Harrison, but through the wealthier, and more populated, neighboring township of Calvin. A native of this area and not of Harrison, Maude remembers, as a child, watching construction crews as they split open the ground of neighboring farms and shaped the terrain into six lanes of concrete, stretching north and south for miles. She recalls listening to the machines work all hours of the night and waking up in the morning sometimes to find that grave robbers had raided a historical graveyard in her backyard, rumored to be more than 150 years old, in order to sell the collectible artifacts one could find. For Maude, what was built (and what was stolen) seemed to have ushered in a gradual change in the area. Because her father's used-vehicle business is positioned on an exit directly off the highway, he has benefited from more customer exposure than he might have had otherwise. At the same time, it was the new highway that brought the massive car-resale lot and the large gas station that currently neighbor her parents' residence.

It was the new highway that attracted County Services to the property across the street from Maude's house for the site of their proposed landfill. The property, a square mile or more of thick forest alongside the highway in Calvin Township, is known locally as "Spaceworld" in reference to a failed development project that was proposed for it a number of years ago. The owner envisioned a massive theme park dedicated to outer-space-related themes, but the project never materialized. In the past decade it had been acquired by County Services. When changing national regulations required CSG to close by 1993, County Services selected the Spaceworld site as the place for its replacement landfill, one that would satisfy the stricter environmental guidelines now in effect. The decision met with swift local resistance. Maude's parents were two of the main organizers behind the effort to stop the landfill, but a number of concerned Calvin residents joined them as well. Eventually, the Calvin Township government became aware of the popular opposition to the proposed development and ceased all negotiation with County Services.[15]

When Calvin Township refused them, the waste company turned, as it had once before, to Harrison. Only a few miles west of the highway, the Harrison border satisfied County Services' requirement for a cooperative township with cheap land. Harrison residents had fought County Services early in that decade when they had attempted to add more acreage to its existing site in the center of the township, which was intended to be a hazardous-waste landfill. Owing in large part to local efforts, substantiated by a report from the Department of Natural Resources, the expansion was prevented. Working with the township board at the time, County Services had to accommodate their latest business plans with the existing politics of preemption in Harrison. Rather than go through the main roads in the northern, more populated half of the township, they proposed to site their new landfill on the edge of town, where it could still take advantage of the highway exit that led in the direction of Spaceworld. A new host agreement was also created, in compliance with state requirements, which promised the township one dollar per ton of waste buried at the site, which, given its impressive size, added up to several million dollars a year—more than Harrison's total town budget at the time.

These were not the reasons given by County Services for the creation of Four Corners. With the new set of environmental criteria imposed by the state and federal governments, County Services was ultimately answerable, not to Harrison or its people, but to state regulators from Lansing and Detroit. And the regulators were far more concerned about Harrison's geological history, which made it ideally suitable for a landfill. As glaciers receded from the interior of Michigan around ten thousand years ago, they left behind thick clay deposits—a natural barrier preventing extensive drainage of materials on the surface into the groundwater below. So, when my employers defended the importation of Canadian waste years later, they would argue that their clients didn't possess the natural inheritance to support landfills of similar scale and capacity. "They don't have the right geology," one of Four Corners' managers would say to inquisitive locals during tours of the site. This might be true of Toronto proper, but the similar geological deposits can be found throughout the Niagara Escarpment, which stretches from southern Michigan through southern Ontario and into Buffalo, New York, partially accounting for the landfills located throughout that region. The truth is that Harrison was fated to become a waste recipient not just because of what it had, but because of what it did not.

Some Harrison residents were relatively complacent about the land deal (as were some, no doubt, in the Calvin area), and it is not difficult to see why, given that many of the township's most prominent politicians and business people at the time were known associates of Bruno and the prominent state politicians who had approved the landfill expansion. In fact, allegations of corruption surrounded the last years of one of Harrison's most prominent and influential political families, partially because of their relationship to the landfill. During his last tenure as supervisor, one of the politicians in the family and his wife were given a place to live (for a reported bargain) in a house on County Services' newly acquired landfill property. This was alleged, alongside other claims of corruption, by local newspapers produced in the neighboring village of Riverside.

The heavily criticized supervisor was eventually ousted in disgrace through a recall campaign. These allegations are not neutral reporting on the history of Harrison, but active and interested constructions with a historicity of their own. The different groups within Harrison did not possess an equivalent means by which to record their own history. It is said that, as they had in 1975, people in Harrison protested the new landfill, however unsuccessfully. Few in Harrison today recall who was involved and what they did. Without a record of their activities, some of the Calvin activists I spoke with are able to insist that Harrison residents did nothing to resist Four Corners. What seems to be the case, arguably, is that activists in both communities did a poor job of communicating and collaborating on their common objectives. According to some, opposition to the landfill in Harrison was concentrated in the predominantly black churches, with which the overwhelmingly white Calvin residents had little or no association. Just as racial politics had segregated migrants on the outskirts of Detroit, so too did they present an obstacle to identification with, and memory of, a more collective struggle.

Through trial and error, County Services was far more proficient at acquiring allies to help them realize their goals. By the time the company began buying up farms in a square-mile plat of Harrison, their plan for a new landfill was already approved at the state and county levels. It was said that the landfill would supply Michigan's municipalities for several decades, thereby avoiding an impending waste crisis. Harrison residents experienced mixed reactions to this rapid and unprecedented change.

Ned Garten remembered how his lifelong neighbors, some of whom could trace back their families in Harrison for generations, became complicit in County Services' bid to take over his family's farmland in the early nineties. He recalled the day his father came home, "pissed off" that the company was seeking to build a new site in their backyard, which would eventually become

Four Corners. Ned was already out of high school at the time and working several odd jobs. At that time, he still had plans to one day farm the area, some six hundred acres of fertile land, as he had with his father and brothers growing up. "This was top-notch land here, you know, the soil was perfect," Ned told me. He loved not only farm labor, but also how people from the area all seemed to know one another and help each other in times of need. It was these same neighbors that were selling off their land one by one, knowing it would become a landfill. Many were older couples, or widows, who no longer wanted the responsibility of looking after a farm, or whose adult children encouraged them to sell. The number of active farms in the township had been declining precipitously since at least the forties, as factory jobs promised more income and less intensive labor. The eighties had been a difficult time for many American farmers and, though people in Michigan's Lower Peninsula did not lose as much as those farther west, there was strong incentive to sell to County Services and move on. Their property had come to seem like a burden, and County Services' offer an opportunity. "They wanted out from underneath it," Ned said, and they "just [saw] dollar signs." Some moved "up north," as many middle-class Michigan residents plan to do one day, while others left the state entirely. The Gartens were not the only family that disapproved of farmers who chose to sell to the landfill. It is probably no coincidence that one of the first houses to sell had all of its windows smashed one night, though Ned believes this was the work of local "Satanists."

The sale of their neighbors' properties presented the Gartens with a dilemma. Ned's parents and grandparents did not want to sell the places where they had lived and farmed for decades, but they relied heavily on share-cropping their neighbors' land in order to make farming viable. Ned's father had long since taken on additional factory work to supplement their crops, which did not produce enough as it was. Within a few months, County Services had acquired a substantial portion of the land the Gartens needed to sustain themselves, and moving became all but inevitable. The Garten family was the last to sell their land, stubbornly refusing until the very end. When they finally realized they had no choice, they waited as long as possible, hoping to get enough money to buy some farmland in a different county, so that they could "do the same thing that they was doin'" for so many years. Ned remembered being unhappy at his family's decision to sell what was effectively his birthright—to be turned into a mountain of garbage. It was with a certain degree of ambivalence, therefore, that he accepted an offer from the landfill to begin working there. This invitation was likely extended partly as a conciliatory gesture to smooth over the transfer of property, and partly out of necessity: "Cause they knew we was farm boys. They wanted a couple farm boys because they were starting a composting operation out there. So [they wanted] somebody that was familiar with tractors and farmin' and what it took, you know, to raise crops."

Ned worked at Four Corners until he lost his job in 2007, though he had managed to advance through the ranks prior to that. It felt strange at first, but not totally dissimilar from farming, "just tearing up the fields and stuff." Tearing up for landfill those same fields he had cultivated as a child, day after day, could not but evoke memories for Ned, surrounded by the loss of both his romanticized past and his imagined future. But chronotopic engagements are not limited to pathos and regret; they involve remaking as much as remembering. Eventually, Ned was made responsible for tearing down his grandfather's house, and his involvement in this taphonomic process was not without its benefits. As difficult as it was for him, in the process he was able to keep and smuggle home, without formal authorization, a large, sturdy beam from the old building. He transported it intact and used it as a central support beam in the barn he was building next to his new home in a new town, near where his grandparents and parents had relocated. For Ned, the beam arguably serves as a memento indexing the trauma of forced displacement from a place where land and memories were once cultivated and shared. Taking and reusing the beam is a material objectification of the place he once farmed and the people he farmed with, as well as the person he had wanted to become. Such chronotopes of emptiness, of lost places and times, are frequently encountered by present and former residents of Harrison.

WASTELAND

After the highway made the area more accessible, Harrison and neighboring townships began to appeal to new groups of newcomers. Between 1990 and 2000, its population increased by almost 10 percent, which was comparable to demographic changes earlier in the century associated with migrating Eastern Europeans, Upland Southerners, and African-Americans. As in the past, these newcomers were received with a mixture of acceptance and ambivalence. However, this occurred at a critical juncture in the history of the township.

As part of Four Corners' host agreement with Harrison, they financed a new, state-of-the-art fire station. Once the landfill began receiving all of Toronto's waste in 2002, furthermore, it began giving the township between two and three million dollars a year in revenue, a substantial amount for a small township with a small tax base. Within a few years, the township became dependent on the landfill for almost half its budget and continued to find new purposes for the money, including an emergency dispatch system, a library, and a recreation center, all of which were intended to reduce its dependency on Riverside. The landfill has become a vital resource, one that residents defend against attack when necessary, as when representatives went to Lansing to stop a ban on the international waste trade, or when dozens of locals signed letters to the state asking them to approve a landfill expansion. At the time I was conducting research, Harrison's acting supervisor did not like the landfill but saw it as necessary. A middle-aged, African-American man, he had moved to Harrison in 1991, after hearing people at his auto plant talking about a place in the countryside (which they incorrectly identified as "Riverside") where people could still get freshly grown food and experience the neighborliness of a small town. At that time, no one had mentioned CSG. By this point, the segregation of Harrison's "Mason-Dixon Line" was a dim memory and African-Americans had moved deep into the southern portion of the township. Many of the attitudes about "newcomers" had begun to change by this time, too, which allowed him to parlay his relationships with his new neighbors into a successful bid for township supervisor, an exceptional accomplishment for an African-American in rural Michigan.

Not all of Harrison's new residents achieved equal acceptance, however. During my time there, some were considered less welcome than Four Corners. A number of newer residents are retired autoworkers. One man I talked with, Harold, had purchased his house to fulfill his dream of life in the country. He had accepted a buyout from one of the Big Three automobile manufacturers and then used the money to finance his large Harrison property. Like many other newcomers, Harold moved to the country so that he could do what he wanted with his land in privacy. But at one of his infamous Halloween barn parties, it was revealed to me by his landscaper—a man named John who had grown up in the area—that Harold had begun feuding with his neighbor, also a newcomer. Though John had been selling pine trees for years, he'd recently been getting more and more work from Harrison residents looking to "block out" their neighbors with nonnative, but locally grown, conifers.

Indeed, some argued that most of the residents who have come in the past decade tend to be less neighborly and to participate less in town civic life. It was said that they tend to think of their residences more as an address or an investment. Some Harrison residents feel the same way about newer businesses, whose owners rarely live in the township. Most controversial, during my research, were a pair of Chaldean party-store owners from the Detroit metro area, Sammy and Ed. They saw Harrison as an excellent location for a small business and had hoped to expand their investments in the township to include a Laundromat. When I would compare these men to immigrants who came to Harrison decades before, one woman of Polish descent shouted at me, "[Polish immigrants] built this country, but these people just want to live off of it!" John, the local landscaper, agreed with this assessment, though I later learned that he and his wife had taken Sammy with them on their most recent vacation to Mexico.

These reactions are partly related to general concern about the direction of the township's future, including how best to raise its profile in the area, how to attract more homeowners and raise the tax base, how the landfill money should be spent, and what should be done when it runs out. Residents disagreed primarily about the future of the town's development—for example, about whether flea markets or more trailer parks should be allowed. During my research, people regularly wrote anonymous letters to the editor of Riverside's newspaper, criticizing the decisions of the planning commission, with introductions such as "The smell in Harrison is getting even worse, like garbage not picked up for weeks, or dead animals." Local politics became increasingly contentious around such issues and began to show up on the Internet. Until relatively recently, a conservative website run by an anonymous resident appeared online, called www.harrisonmorons.com. Aside from posting a new "Moron of the Month" page lambasting local politicians, the site also included a "blight page" with photos of properties considered poorly kept by residents, as if policing the landscape for people and homes that did not belong.

As the question of "what ought to happen" in Harrison turned into "who doesn't belong" or "who is hurting the township," stories began to circulate about wild animals. Roaming packs of wild dogs had long been a matter of concern. These dogs, it was widely believed, were unwanted strays that had been dropped off by people from nearby towns and suburbs. If correct, this would mean that pet abandoners perceived Harrison as a rural threshold beyond their familiar surroundings. Another possibility is that the dogs wandered in of their own volition, perhaps attracted by the lack of people and the smells of waste. Whatever the reasons, the dog packs were an acknowledged problem. Some residents adopted the dogs they found, as those living in the vicinity of Four Corners had done in the past, or took them to shelters, as Phyllis recalled doing while growing up. If not recovered, these strays might form packs. To survive outside of human companionship, they had been known to scavenge food from trashcans, or to stalk and attack small pets, chickens, and even ponies. They were regularly spotted in the vicinity of the landfill at night, moreover, and could be identified by eyes glowing in the distant darkness. Dog packs were considered enough of a threat that Harrison police would shoot them on sight; controversially, they did not allow others to do the same. If their purported origins can be believed, the dogs were a mobile and dangerous reminder of the consequences of Harrison's representation as empty countryside.

Other animal waste and wasteful animals demonstrate Harrison's peculiar fate as a destination for people and things out of place. During deer season in rural locales, it is not uncommon to find the stripped carcasses of hunted animals that were left by the roadside to decompose. The management at Four Corners have come to expect such deposits around their perimeter every year, though they do not clean up the animals themselves. To my knowledge, these carcasses have never been much discussed by locals. Like the operations of the landfill as a whole, such illicit dumping usually occurs out of sight. Far more troubling for many residents is the one-hundred-pound female cougar that was periodically spotted in the southern portion of the township during 2004 and again in 2006. Although the Department of Natural Resources has reported that cougars have been extinct in Michigan for more than a century, raising the suspicion that these reported sightings were hoaxes, many locals claim to have seen the animal. Others have found their livestock dead, necks broken, alongside the tracks of a large feline. Like the wild dog packs, the cougar's presence suggests a naturalization of political anxieties concerning inside and outside, wild and developed, that have been at the center of political debate concerning the future of the township. More than political symbols, these destructive creatures quite literally give bite to territorial divisions of labor.

It is somewhat ironic that many of those seeking "country living at its finest" (Harrison's official motto) are disturbed by the presence of wild creatures and anxious to begin developing the township. In another sense, the cougar and the wild dogs make Harrison appear to be a sort of wasteland for unwanted and feared creatures, just as it has become a dumping site for foreign refuse. Harrison is now a place where people go to hunt deer and a place to recycle them on the roadside, a regulated wilderness area and a place to let one's former pets run wild, a dumping ground and an appealing countryside. Its littered landscape expresses many of the contradictions that dominate rural North America because of the fantasies that it inspires in others.

BEHOLDEN TO ABSENCE

What happens when a powerful city, like Toronto, sends its waste far away, to places few people have heard of? If I had not used pseudonyms like "Harrison" and "Brandes" for the places in this book, few people would have recognized their actual names. In a way, that's the whole point. Imbalanced exchanges like this allow people to imagine that a city's growth knows no material limits. It is as if money can be made, buildings rebuilt and resurrected, life lived without concern for the real conditions and limitations of the material world. And the apparent transcendence of Toronto is no mere fantasy—eventually, none of these places will exist, but people in the future will know far more about Toronto, for far longer, than about any of the places it sent its rubbish. The translocal movement of waste not only presupposes such a hierarchical division between places; it helps to maintain it taphonomically.

Staying in the same place can be just as hard as moving someplace new, and these are not new problems. Thousands of years ago, Michigan's indigenous inhabitants smelted precious copper; some of it was kept and some of it transported to the Southeast Atlantic Coast. In what is now known as Georgia, it was ceremonially disposed of, with a mixture of animal and human remains. According to one hypothesis, coastal dwellers did this in order to facilitate the permanent settlement of villages for the first time (Sanger 2015). Waste deposits consisting of corpses and rare valuables may have served as a form of collective memory, a monumental burial that could provide a meaningful anchor for lasting occupation. Centuries of settler colonialism in North America—fueled in part by representations of the continent as empty and by a renewed extraction of ores from Michigan's Upper Peninsula—have radically transformed the contemporary movement of people and valuables. It is now Michigan that has a crisis in settlement. First, people moved out of the old industrial cities, the precious metals of which are now stripped and resold to travel the world as scrap. The mass waste that crosses the border from other states and Canada does not help resolve Michigan's dilemma of de-settlement. And, in some cases, waste odors and traffic have reduced housing values, creating difficulties for Michigan homeowners who increasingly seek to leave the Rustbelt behind for the Sunbelt of the southeastern coast and southwestern desert.

We are beholden to absence as it pulls on us in two directions. On one hand, the past exists no longer but continues to shape the present; on the other hand, we collectively anticipate and shape a future that does not yet exist, in relation both to the dead that came before us and to the dead we will become.[16] Harrison's rural historicity bears traces of its complex relations with more populous neighboring communities and more powerful national centers—which have influenced its development and filled its landfills. As a consequence, the past and present preemption of the township by and for outsiders cannot help but weigh upon those responsible for its uncertain future. Then as now, residents of Harrison recognize that what really distinguishes their home from neighboring areas is its natural surroundings. As we have seen, the appearance of an untouched, rural environment is itself a product of socio-natural history. The absent traces of that history, their concealment and subtraction, are what help make it seem rural and, for some, an appealing place to start over. Just as this continues to attract new residents to the area, it keeps long-time residents from moving. It is also the reason why some wish for the landfill to remain in business and why others want to see it close down.

Ned's loss of his family's land and his involvement in its transformation into something new represents well the ambivalent position of many Harrison residents who, in different ways, tend to recognize the preemption of the township's landscape and the replacement of prior meanings and memories with uncertainty. Ned's story also demonstrates, along with the many other developments, schemes, and fantasies I have mentioned, the close connection between the peculiar socioeconomic fortunes of rural areas and the historicity of their landscapes. For Ned and Phyllis, the Harrison they knew is being lost as it is more and more shaped by the interests of the world outside. This is nowhere more apparent than in the way different corporate and government actors have intervened over the years to procure Harrison's land for the waste of other places.

One cannot separate the township's current fate from the broader dynamics that have shaped its transformation into a rural wasteland, which both preceded and made possible its landfills. The different absences associated with these multiscalar relationships help account for Harrison's history, but they also highlight the partiality and fragility of its historicity. Partly as a result of the area's uncertain past, residents now struggle over an uncertain future, one with and without Four Corners landfill.

FIVE

_____

Ghostly and Fleshly Lines

ACCORDING TO CHAPTER 3, some waste is better than other waste, depending on whether it can be remade into something worthwhile—and, in turn, remake people and their relationships. By implication, some kinds of waste are also worse than others. This is codified in North American and transnational waste regulations, which make distinctions between waste streams according to their relative hazardousness and the labor and technology required to manage them. Nothing is merely "waste in general." Governments and industries distinguish between waste that is more or less expensive to dispose of, individuals and communities distinguish between more or less dangerous waste to work and live with, and waste scavengers distinguish between more or less valuable waste to recover.

The category of "toxic waste" has been particularly important for the evolution of environmental regulations and supranational agreements over the past half century. The dramatic growth of the U.S. economy after World War II relied heavily on chemical manufacturing, which gradually raised concerns about the novel forms of pollution associated with it. By the 1980s, environmentalists had succeeded in both problematizing the disposal of hazardous waste and promoting recycling as a more sustainable alternative to the disposal of nonhazardous waste. At the same time that new regulations came into effect concerning proper treatment and disposal of distinct waste streams in North America and elsewhere, long-distance transport to and from these places was becoming cheaper and more efficient. As a result, regulatory asymmetries and low transport costs introduced an expanded global market in waste services, as happened in the Great Lakes region on a smaller scale.[1]

With all kinds of waste moving far from their places of origin, a new way to distinguish them has emerged. Waste can now be additionally categorized on the basis of where it came from and who produced it, as separate from whatever qualities, good or bad, that it may possess. Some types of waste are now considered worse than others, not necessarily because they are objectively or subjectively so, but because they stand in for social identity and difference.

This was true of the Canadian waste that was landfilled in Michigan from 2002 to 2010. In 2005, near the peak of this activity, Michigan imported approximately six million tons of waste from out of state, two-thirds of it from Canada. This was 18.6 percent of the total amount of municipal solid waste disposed of in Michigan. It was 13 percent of all of the municipal solid waste imported by U.S. states that year.[2] The Canadian waste trade exposed my coworkers to public scrutiny and criticism. Big Daddy, generally unaffected by the social stigma of dirty work, echoed the opinions of others when he expressed how he had been affected by the unpopularity of the waste trade:

> Everybody that I've run into and my wife runs into . . . with her job and when they find out that I work down here and we take Canadian trash, automatically everyone's against that. My method is I don't tell people where I work. Cause I don't want to get into an argument. If people ask me what I do I just tell them that I'm a construction supervisor.

For Big Daddy, this was about more than mere embarrassment:

> I always feel like, you know, who else in the state of Michigan can get up in the morning and go to work knowing that the whole state hates you: the governor hates you, the Wayne County executive hates you, the people hate ya. Who who who does that? Very few people in the state of Michigan. Most people when they get up in the morning and and get their cereal and their cup of coffee and they're getting ready to go to work they don't go to work thinking that the whole state hates them because of political and emotional issues, you know, probably not. They know that the whole state don't hate 'em, where you know that . . . here.

Big Daddy felt vilified for corporate and political decisions beyond his control, and he was not alone. Some claimed that they were not bothered by this general animosity, while others angrily referred to public dislike as a nuisance. One white Canadian truck driver joked that other motorists were constantly telling him he was "number one," raising up his middle finger to me with a grin. Some landfill workers and waste haulers expressed sympathy with those who opposed transnational waste imports but, like Big Daddy, felt unjustly targeted for trying to make a living.

The imagery in Big Daddy's words ("get up . . . get their cereal and their cup of coffee . . .") reveals a desire to be recognized as normal, ostensibly like all of Michigan's employed residents, satisfied with the anonymity of their daily routine. Doing a job well will never be enough, so long as others deprive him of the right to take pride in who he is through what he does. My coworkers occasionally joked about popular disdain for Canadian waste. I recounted one such instance in my field notes:

> I am picking the road with Eddy and he remarks, "There's nothing worse than a maxi pad covered in dirt so you don't know what it is until you pick it up." I agree with him. "Know what's worse?" he continues, "Canadian truck driver maxi pads!" [Laughter]

From the perspective of a laborer, it may seem absurd to distinguish the quality of waste on the basis of national origin. But for many Michiganders, the importation and burial of Canadian waste was considered a violation of American air space. One of the local activists I met, named Jacob, placed a sign in his yard along the roadside that read simply "CANADA—1 MILE AHEAD." For Jacob and his friends, it was as if Canada were exerting sovereign control over their neighborhoods through the aegis of a free market in waste disposal. That Michigan was being trashed by a foreign government seemed unacceptable and unreal. For critics, it was self-evident that waste could belong to Canadians and therefore *belong in* Canada.

One of the most outspoken opponents of the Canada–Michigan waste trade that I met was Bill, a fifty-seven-year-old small-business owner who was instrumental in organizing small demonstrations against it in 2006. Bill made it clear to me that he was opposed not to Four Corners—as many landfill managers and employees assumed—but to Canadian waste in particular. I asked whether residents of Calvin Township were equally upset about waste coming from other towns, counties, or states as they were about the waste from Canada:

BILL: I think people are all right with our own garbage.

J. R.: Why do you think that is? Because it could just as easily stink.

BILL: Well here's a perfect example of that—let's just both get over here and I'll shit on the middle of the floor and you shit on the middle of the floor, and if you have to clean up both piles would one be more offensive to you, Josh? You know, if I shit in the middle of my floor I don't like it, but it's my shit!

The shocking nature of this imagined act, the fact that it combines moral inappropriateness and physical repugnance, demonstrates Bill's intense feelings about the harm and injustice caused by Canadian waste entering Michigan. At the same time, it illustrates a more basic belief: waste is fundamentally bound to the people and polities that produce it, like a physical trace one leaves behind.

It is not so strange that people should be identified with their circulating waste. Many nonhuman animals leave traces of themselves around their habitats, in the form of scat or scent, to mark their territory. In a sense, they are laying claim to their environment by leaving a part of themselves behind to be interpreted as a sign of their growth and ongoing occupancy. Waste is deposited as a signal to keep rivals at bay or to attract familiars; it is a marker of identity and difference that separates one animal from another. A cat's urine is not merely matter out of place according to a human classification of the world, for instance, but an intraspecies form of communication important for feline (a)sociality.[3] But the direct spatiotemporal connection that cats, Bill, and I share with our specific waste deposits is very different from the symbolic associations that arise as a result of the shipment of waste across national borders. National lines are both more enduring and more ethereal. The label "Canadian" can stick to something no matter the actual qualities it possesses. All the waste loads from Toronto might be considered "Canadian," even though tourists, migrant workers, and transnational citizens from other places also contribute to the mass waste of that global city.

The "Canadianness" or "Americanness" dividing one person, thing, or place from another is difficult to reveal in a directly perceptible way—there is something ghostly and imaginary about this kind of identity and difference.[4] Ghostly lines have to be performed as though they were more obvious and perceptible, which is what Bill did in calling upon his and my polluting bodies. Bill's conception of proper waste management—for individuals or nations— as taking responsibility for "one's own waste" is a sentiment many other Michigan residents share. It is not for governments, businesses, or markets to alter the appropriate destination of waste, because all waste products are morally bound to those (national collectivities) that produce them. As Bill put it:

> I just think, personally, I'm all right with Michigan garbage. It is the principle that each and every one of us have got to take care of our own shit. If a state produces an amount of garbage a year they should be able to control their own garbage. It goes right down to my personal stuff: I think we should take care of our own stuff.

The adjective "our" presumes a "their," presupposing a further divide between the us/them of nation-states. Bill went on to explain why Canada was particularly irresponsible with its own stuff, according to his perspective: "When someone has the resources like Canada does—they're Siberia! You don't have to go very far north of Toronto and it is barren. And that, I guess, is what upsets me more."

Equal parts direct, figurative, crass, and practical, Bill's analysis is indicative of the complex and shifting affiliations and boundaries that undergird waste politics in the world. In some ways, this is analogous to the shift in meaning that us/them and mine/yours undergo in varying contexts of use. On one hand, neoliberal ecogovernance hinges on making individual citizens responsible for the health and sustainability of their surroundings. Like many North Americans, Bill and his fellow protesters have been targeted by numerous initiatives to cultivate their sense of individual responsibility for environmental problems like littering, recycling, dwindling fossil fuels, and climate change. Bill's conviction that "taking care" is a "personal" task reflects this importance placed on individual conduct. On the other hand, Bill offers a challenge to this view by suggesting that, like individuals, states have obligations too.

And yet, paradoxically, when polities end up exporting or importing waste, it is precisely to fulfill these obligations. States may inadvertently promote waste importation while officially trying to "take care of our own stuff." In the 1980s, Michigan encouraged the siting of new landfills in order to ensure enough dumping space for the future. Each county was required to develop a waste management plan, given the available air space of their existing landfills. As an incentive for waste firms to build landfills in their state, the government offered tax-free bonds toward waste facilities. With a surfeit of landfill space, some large enough to handle local waste-disposal needs for decades into the future, landfills had to compete more in order to get local contracts. In search of profits, Michigan waste companies could offer lower landfill tipping fees than their competitors in neighboring states and provinces. As some local environmental groups anticipated, Michigan's plan resulted in an expanded regional market in waste services. While the ability to increase landfill capacity is controlled by state and county governments, moreover, the authority to regulate imported waste belongs to Congress alone, as provided by the Commerce Clause of the U.S. Constitution. This prevented Michigan politicians and activists from controlling the waste market once it was introduced.

One could argue, as my former coworkers at the landfill tended to, that Bill's appeal to individual and national obligations was merely a selfish, NIMBY-like tactic to resist the landfill's presence near his home. In this chapter, I take Bill at his word that it was the imposition of Canadian waste that he resented in particular. According to Bill, even as mass waste is alienated from the people and places that produce it—as abstract negative value—it remains their collective responsibility. Yet, as I have argued throughout this book, that association is not guaranteed. For local activists like Bill, politicizing waste as Canadian meant performing their own victimization and forming alliances with people who could make it apparent to others. The vanishing connection between waste producers and products, which Bill found self-evident, can be politicized only by drawing attention to the taken-for-granted circulation of waste, whereby some people and places assume responsibility for transient matter on behalf of others. It is necessary, in other words, to performatively enact the lines of division implicated in waste circulation. It is for this reason that Bill associated the ghostly line dividing Canadian/Our waste with a more concrete distinction: the actual traces that our bodies more directly leave behind (Your shit/My shit). This analogy helped Bill make the ghostly lines of his political position more meaningful and concrete.

The concreteness of bodies, their carnality, makes them powerful vehicles for expressing social difference. Nativist boycotts against foreign commodities in Michigan, for instance, have occasionally appealed to enfleshed racial divides to gain more purchase. The racially marked bodies of Asian-Americans were subject to structural and real violence when the Buy American initiatives began to take hold in the Midwest and nationwide. The image of the color line was invoked by Frederick Douglass and W. E. B. DuBois to depict the prejudice and structural violence affecting people in North America and abroad. Like that of borderlines, the naturalness of these boundaries is imagined—yet both are made concrete, naturalized in practice.[5] The entangled enfleshment of national and racial lines, their imposition on the lived realities of those so marked, makes them seem less ghostly and abstract. This chapter will explore the relationship between the attempted politicization of waste circulation and the ghostly *cum* fleshly lines of nation and race implicated in efforts to imagine and contest it.

The orchestrated movement of transient materials elsewhere is implicated in environmental racism throughout North America, and (anti-)racist resistance can impact waste markets in unexpected ways. Unbeknownst to Bill, the waste loads that ended up going to Michigan landfills, including Four

Corners, were originally meant to go to northern Ontario, just as he suggested. The site that was chosen was an abandoned mine located within Timiskaming ancestral lands at the border between Ontario and Quebec. For the indigenous Timiskaming First Nation and their allies, that waste was marked not as Canadian, but rather as southern, nonindigenous, and urban. It was on the basis of these lines of division that they organized successful resistance to the planned landfill. It was only when that waste left Ontario entirely, bound for Michigan, that it became Canadian.

Just as nationalism and racialization shape the distribution of environmental harms, moreover, so too do they shape the articulation of new political movements. Attempts to reconnect abstract negative value with the places and people to which it "belongs" must fix a social identity to anonymous mass waste. When it came to Four Corners, this personalization of substance brought with it a substantialization of persons and the ghostly lines that separate them. When marked bodies become mobile carriers of difference, national borders may be more concretely reenacted. Both my former coworkers and local residents living near our landfill were invested in rhetorics of whiteness and tended to employ local repertoires of race in their encounters with incoming waste and truck drivers. In this way, the ever present possibility that anything could be in the waste was associated with the threat of foreign and minority truck drivers of unfamiliar background and unknown motivation.

RACIAL ECONOMY AND WASTE CIRCULATION

As discussed in chapter 2, capitalism encourages a dominant interpretive tendency that elevates the value-form of money as the primary motivation and explanation for social action. Many things are readily interpreted in this way in North America, but when waste circulation is solely determined by the market forces of supply and demand, the result can appear obscene. This was the general reaction to a leaked memo attributed to Lawrence Summers when he was still chief economist at the World Bank:

> [S]houldn't the World Bank be encouraging *more* migration of the dirty industries to the LDC's [less developed countries]?. . . I think the logic behind dumping a load of toxic waste in the lowest wage country is impeccable and we should face up to that. . . . I've always thought that under-populated countries in Africa are vastly *under*-polluted. (quoted in Clapp 2001: 1)

This memo was reprinted in *The Economist* in 1992 and created immediate controversy, leading to an eventual apology from Summers. Even if one were to accept his claim that the statement was meant sarcastically, the timing was inauspicious. The memo's release co-occurred with the introduction of the Basel Convention, a new international agreement prohibiting the trans-boundary movement of toxic waste from rich to poor countries. The Basel Convention did not stop the practice of transnational dumping, but it did raise global awareness of the wealth disparities that made some parts of the world likely destinations for waste. The Summers memo would have had waste move as an anonymous bad, unimpeded by its connection with those who produced it and the obligations ensuing therefrom. If the anonymity of waste flows was instead revoked, their connection to specific people and places made visible, then those people and places would become newly accountable for where their waste ends up.[6]

The Summers memo may have been widely criticized, but the representation of poorer people and places as somehow more deserving of pollution has not gone away. To imagine Africa as underpolluted is to ignore how the disparities in health and regulatory infrastructure—which make dumping economically attractive in the first place—would also make the same waste *more harmful* to Africans than to their wealthier counterparts in the Global North (Swaney 1994). The same misjudgments about Africa, as an abstraction, are behind both relative underinvestment in African economies and the normative expectation that it is a continent where environmental harms (rather than investment capital) should go (Ferguson 2006). In other words, unreal abstractions have real impacts—without them, actual places and people would suffer less. This is something the ghostly *cum* fleshly lines of race and nation have in common. At the same time, the divisions they create between people are not natural, so they must be continually reenacted and made concrete so that they seem that way.

The various social processes and practices that reproduce racial discrimination make up a racial economy, one that systematically racializes our disproportionate subjection to suffering and access to privilege.[7] This includes, notably, disproportionate exposure to environmental harm. In postwar Detroit, for example, African-American city workers were most often relegated to filthy and unskilled work; and in the United States generally, dirty trades like scrap recycling and junkyards have historically been controlled by ethno-racial minorities.[8] The involvement of racialized groups with dirty work further marginalizes them through their association with polluting activities. And though many waste entrepreneurs were dispossessed through corporate takeovers during the course of the twentieth century, there is a lingering tendency to associate waste management with ethnically marked whites and organized crime. As discussed in chapter 4, throughout North America, the idea of wasted property was used to dispossess indigenous people of their lands. The idea of wasteland continues to serve as a rhetorical trope in the unequal distribution of risks to indigenous communities in North America, as if the places where they dwell were empty and underpolluted.[9]

The relationship between racial economy and uneven waste distribution gained attention and influence in the 1990s. Environmental-justice scholars have provided evidence of the disproportionate impact of pollution on the poor and communities of color and documented the efforts of social activists and transnational movements seeking recognition and reform.[10] It is also worth noting where transnational networks do not form and possibilities for political action go unfulfilled. The possibilities for recognition and politicization are different precisely because waste circulation is differently enmeshed in distinct histories and practices of social difference.

When city officials from Toronto originally chose to use an abandoned mine in northern Ontario as a replacement for the landfill located just outside the city, they did not seem to have anticipated much resistance. They were, after all, keeping waste within the province, thereby implicitly fulfilling a commitment to at least one geopolitical us/them distinction. The city's historical landfill was due to close, and they knew they would soon need a replacement. Siting and expanding waste-disposal facilities amid the growing residential neighborhoods of the greater Toronto area would be next to impossible. This meant that a distant, rural locale was more feasible. But the mine had been controversial when it was still in operation and was now a reminder of an era when extractive, southern industries encroached on indigenous lands and communities. In the early 1990s, the Timiskaming First Nation began organizing to protest the proposed landfill, symbolized as a resumption of neocolonial environmental abuse.

Indigenous resistance to Toronto's new landfill can be better understood in the context of the North American racial economy. From the 1960s onward, indigenous Americans began organizing politically for recognition and self-determination. Although various indigenous struggles had been ongoing for centuries, members of indigenous communities who were newly educated or returning from military service helped mobilize resistance to racialized structural violence as part of a more visible and organized movement. In Canada, indigenous resistance tended to be particularly focused on high-profile meg-aprojects that threatened First Nation ways of life. The James Bay Cree of Quebec, neighbors to the Timiskaming, had spent decades struggling with the consequences of hydroelectric development. In the 1980s, the failure of provincial and federal governments to provide agreed-upon services and infra-structure was linked to the tragic death of seven Cree children, the apparent victims of inadequate sanitation.[11] Since Timiskaming were likely aware of these highly publicized struggles, there was no reason for them to expect that their rights would be respected if Toronto's waste began to come north. The struggle for indigenous self-determination does not stop with the problem of environmental injustice.[12] However, when indigenous communities are disproportionately affected by the allocation of social bads, this brings the structural violence of racial economy into relief.

Confronted with a decade of organized political action by native and non-native activists, coupled with the official closing of Toronto's landfill, city offi-cials finally signed a contract with Four Corners. After a few years, the amount of transnational waste landfilled in Michigan would be close to the largest recorded in the world at that time. It was, in a sense, a fulfillment of Summers's vision. The North American Free Trade Agreement (NAFTA) forbade protec-tionist restrictions on transborder economic exchange, including waste serv-ices. Along with the Commerce Clause of the U.S. Constitution, NAFTA freed Ontario's waste to become purely abstract negative exchange value, unmoored by connection to any polity or population, dependent entirely on the marketized supply of, and demand for, pollutable space.

Despite the clear disparity in wealth and power between Toronto and Detroit, the waste trade between them did not attract the attention of inter-national organizations like Basel Action Now or Greenpeace. According to the nation-centered discourse of global waste injustice, the flow of waste from southern Ontario to southeastern Michigan would have been classified as a transaction between relatively wealthy nations. Arguably, this invisibility on a transnational stage limited the means that Michigan residents had at their disposal to frame their growing discontent.

MAKING WASTE CANADIAN

People from southeastern Michigan are accustomed to representing com-modities and politicizing foreign trade as polluting or destructive. Today known as the "Rust Belt," the manufacturing sectors of the Great Lakes region began to suffer from massive layoffs and economic precarity in the 1970s. Since then, national divisions have been closely associated with a sense of individual responsibility through the widespread Buy American campaign, a tactic that has successfully mobilized consumer-oriented political action in the United States for centuries (Frank 1999). By the early twenty-first century, these initiatives had become common sense for many Michiganders and a moral guide for acts of consumption. Though none of my coworkers at the landfill worked for the auto industry (and quite a few were entirely unsympathetic to the plight of auto unions), I was still occasionally chided for driving a Toyota to work.

The politicization of waste as *Canadian* comes easier in some ways, precisely because so many people already value mass waste negatively as something they reject in their everyday lives. At the same time, the relative invisibility of waste circulation and disposal complicates its transformation into an arbiter of identity and difference. In contrast to the conspicuous circulation of automobiles and the very public display of make and model worn on their exterior, for example, waste hauled into Michigan is hidden from sight and its national origins may be difficult to discern. Located at one of the busiest borders in the world, the actual passages between Detroit and Windsor (or Port Huron and Sarnia) concentrate attention on the ghostly line that divides two powerful nations. Canadian waste haulers witnessed this performance of national difference every day and became accustomed to it. Once through the border checkpoint, they might maintain anonymity until they reached their destination, much like their anonymous haul. Not many motorists knew what these trucks carried, though they might detect a peculiar odor if caught behind them. It wasn't until waste haulers reached the county road leading from the highway to Four Corners that an us/them performance of national lines reappeared.

In general, waste haulers from Canada were under greater local scrutiny than their counterparts, and they knew that. As they took the highway exit leading to Four Corners, waste loads hauled from Toronto became marked as Canadian, as were the drivers. Those residents who wanted to voice their opposition to the Canadian waste trade placed signs in their yards along the county road, for passing drivers to see. Some of these signs had been provided by political parties, and others were homemade. Though it had begun in the 1990s, the movement of waste across the U.S.–Canadian border did not receive widespread public attention until the City of Toronto began shipping

TABLE 1 Timeline of legislation and local activism associated with the waste trade

2003	Wayne County proposes an amendment to its 1999 Solid Waste Management Ordinance that would ban waste imports that do not conform to the 1976 Michigan Bottle Bill.
	The waste industry files suit, claiming the amendment is unconstitutional.
2004	A federal court finds against Wayne County's proposed amendment in *Waste Management Holdings v. Wayne County.*
	Odors are reported in the vicinity of Four Corners during spring and summer.
	The "Don't Trash Michigan" campaign distributes signs along the county road.
	John Kerry campaigns for president in Michigan, claiming he will seek a ban on waste imports.
	Governor Granholm signs laws requiring imported waste to meet Michigan standards (this has no effect after waste imports are proved to be in compliance).
2005	The "trash-o-meter" becomes a popular website that purports to show how much out-of-state waste has entered Michigan since the beginning of 2005.
	Granholm and state Democrats propose raising the tipping fee from 27 cents per ton to $7.50; this is rejected by the state legislature's Republican majority.
	Michigan passes House Bill 4760 to ban further landfill expansion; Four Corners applies to expand before the law comes into effect.
March 2006	A sludge spill occurs on the county road.
	The first demonstration of local activists occurs on the county road.
	The Michigan Democratic Party coordinates a rally at a wetland preserve with state legislators.
	A town hall meeting is held at the Calvin Township senior center.
April 2006	A protest is held at the capitol building in Lansing.
	The Michigan Department of Environmental Quality holds a hearing in Brandes to consider Four Corners' proposed expansion.
May 2006	Four Corners stops taking sludge, a decision that proves temporary.
	A lawsuit is filed against Four Corners by local residents.

all of its waste to Michigan in 2002. Prior to her election as Michigan governor, Jennifer Granholm had been criticizing Toronto officials since early 2001, when Michigan rose above Virginia as the nation's second leading waste importer. In 2004, the same year that Governor Granholm signed an unsuccessful bill that was meant to stop the flow of Canadian waste, the county road was decorated with signs in support of the "Don't Trash Michigan" campaign. Despite these and other attempts by political representatives and activists (see table 1), the amount of waste imported from

FIGURE 7. Local protesters.

Canada steadily increased, from just over 2 million tons in 2002 to over 3.5 million in 2005, nearly 20 percent of the waste produced in the state of Michigan that year.[13]

I first encountered people demonstrating against Canadian waste importation one morning in the spring of 2006. Only days before the demonstration, a Canadian truck had gone off the road and spilled several gallons of sewage sludge, not far from where they lived. The road was shut down for the entire day, and traffic to the landfill was rerouted through the middle of Brandes.

On the day of the demonstration, the protesters stood at the intersection of Brandes Street and the county road with large handmade signs, three middle-aged men and a younger man, all of them white, distributing flyers to passing motorists about an upcoming meeting. Every few seconds, they hollered "Go back home!" to passing waste haulers. I introduced myself and discovered that, despite what many landfill employees believed, they were from the surrounding area and not Lansing or Ann Arbor. The men had known each other for years and were members of the same local church. The three older men, Bill, Jacob, and Roy, were fairly prominent in the politics and businesses of Brandes, a small village in Calvin Township. I also learned that there was a fifth demonstrator, Ron, a middle-aged, white, small-business owner who lived farther north in Newton. He drove by as we were talking and was greeted with cheers from the other demonstrators. His pickup truck was moving at a slow crawl down the fifty-mile-an-hour county road, stranding tractor trailers behind him as they headed for the border to pick up another load of Canadian waste.

The demonstrators told me that they wanted to draw public attention to the waste trade and what it was doing to their community and their lives. According to Bill, they also hoped to affect the landfill's business in some way and force its owners to recognize the harm they were causing (that was why Ron was trying to cause a traffic jam). But the decision to take political action did not come easily to them. As Bill would later tell me, "I'll be honest with you. In my younger days I woulda took a activist and strangled him and put him up in a tree." While Bill did not say so explicitly, this comment could be taken as an oblique reference to the lynchings and white-supremacist backlash against the civil rights movement of Bill's youth (below, I'll discuss the role of racialization and white privilege in the political tactics of these activists). Many of those opposed to Four Corners, though by no means all, were staunch conservatives who suspected "activists" of being people who didn't have real jobs or responsibilities and wanted to interfere in other people's business. Their decision to take action came after thirteen years of living down the road and downwind from Four Corners. The recent spill was only the most recent provocation; they claimed that the odors were getting worse and reaching farther distances, even all the way to Ron's house.

These criticisms were not a result of Canadian waste exclusively, but were part of the cost of living down the road and downwind from an operating landfill. Yet the demonstration was explicitly directed at the transnational waste trade, and all those involved claimed that the location of the landfill was less of a concern by comparison. The signs they'd made the night before read "No More Trash from Canada," "Keep Canadian Trash in Canada!!," and "No More Crap from Canada, Take It Back to Ontario." Partly, this approach was adopted to attract broad support for what were problems specific to their own neighborhoods. The waste trade had been documented statewide by environmental groups, politicians, and journalists. Many Michigan residents saw Canadian waste as a danger to their state's environment and, in a more general sense, its future well-being. The southeastern Michigan residents I knew found it particularly troubling that a state known for its freshwater resources and picturesque Upper Peninsula became the preferred dumping ground for a foreign country. "This is supposed to be the Great Lakes State," I was often told, "not the Great Waste State!"

For many Michigan residents, waste trafficking across the Canadian border took center stage as a meaningful trope for widespread fears about personal, economic, and environmental insecurity in a world dominated by challenging new forms of global connection. International environmental controversies are particularly well suited to channel border anxieties because they have the potential to give the nation-state's inclusions and exclusions the appearance of practical necessity and ecological urgency, which gives the appearance of unifying divided publics behind a common cause. Discussion of Canadian waste would arise in tandem with state elections every few years, each time characterized as an urgent problem of the moment. Various websites, radio programs, television stories, and newspaper articles were devoted to it. Canadian waste was an important factor in several election campaigns in Michigan, particularly for Democrats, and was the subject of a slew of bipartisan bills passed in the state legislature, six bills in the U.S. House of Representatives and two in the Senate, all seeking to regulate or abolish international waste imports through different means. In a sense, opposition to the trade seemed to lie beyond politics altogether, insofar as there was little disagreement *that* it should end, only disagreement *how*. No vocal defenders of the trade existed from either political party.

By linking their demonstration to the transnational waste trade, the local activists I met were being strategic but also trying to give their struggle meaning. In this way, they hoped to channel existing opposition to waste imports to effect local change. The staging area for many of their activities was Lions Service, the small party store and towing operation at the center of town, which a handful of local families frequented. Maude and Gwen, who owned the business, had been actively involved fifteen years prior, when the owners of County Sand and Gravel had sought to purchase land across the street from members of their family.

Despite their recruiting efforts, the number of activists regularly involved in local demonstrations and rallies never reached more than a dozen or so, which may be partly due to their greatest strength—their deep and long-standing connections to the local community. Most of the other vocal opponents of the landfill were newer residents living in the subdivisions, who tended to avoid places and events that seemed designed for community insiders. The activists accused their neighbors of being too caught up in their own families and careers to make time for local politics (not unlike the sentiments of Harrison residents toward recent newcomers, mentioned in chapter 4). They also conceded that the landfill tends to affect them more because of the location of many of their residences along the county road.

Ambivalent Alliances

Prior to the demonstrations against Four Corners, the Michigan Department of Environmental Quality (MDEQ) published notices, in several local newspapers, of America Waste's intention to expand Four Corners into fifteen additional acres of woodland, extending their already sizable dumping capacity by approximately 3.2 years. The sludge spill on the county road that sparked Bill, Jacob, and their friends into action had come only weeks before the expansion was to be brought before the public. As with many landfills throughout the state, the application for the planned expansion was submitted to the MDEQ at the very end of 2005, right before a legislated moratorium on landfill expansion came into effect, which was intended to curtail the transnational waste trade.

In response, the activists affiliated with Lions Service formed new political alliances. Their initial demonstration had attracted some media coverage and drew the attention of the Michigan Democratic Party (MDP). The MDP had organized local campaigns against Four Corners before, using it as an emblem of the unpopular Canadian waste trade during the previous gubernatorial election. Anticipating the pivotal Senate and House races of 2006, the plight of Brandes residents offered the MDP an opportunity to revisit a well-worn political issue that had a broad base of public support. Brandes residents, particularly those affiliated with Lions Service, did not mind being used by politicians for this purpose since they felt strongly that the Canadian waste trade was largely responsible for their tribulations. While they were ambivalent about contributing to partisan election-year strategies, they were willing to use politicians to help publicize their struggle.

This mutually beneficial collaboration led to two local rallies and one political demonstration at the capitol building in Lansing. The first took place in late March, an impromptu political rally in the wetland preserve beside Four Corners, some weeks after the initial demonstration on the county road. The rally was arranged by the Democratic Committee on behalf of two representatives running for reelection in the House. Congresswoman Rachael Paterno, the state representative for both Harrison and Brandes, had been a vocal opponent of the waste trade for years and had earned a great deal of support in her district as a consequence. She was also participating in a town hall meeting at the Calvin Township senior center the following week. Flyers for the event displayed her name prominently, followed by the words "Stop Trash," "fighting landfill expansion," and "dealing with offensive garbage odor." Paterno's colleague at the rallies, Representative Krysten Zimmler, grew up not far from Maude's childhood home and represented other districts near Four Corners.

Ostensibly, the rally and the town hall meeting were meant to bolster support for measures in the legislature that would sustain the statewide moratorium on landfill expansion and discourage any further addition of air space from attracting more out-of-state waste. More importantly, the rally was meant to increase local involvement in the MDEQ hearing to be held at Brandes High School in mid-April to consider the proposed expansion. The state representatives and the MDP hoped that the rally and the town meeting would create momentum leading up to the local hearing, which would provide a large enough audience to influence both state lawmakers and the MDEQ.

The rally was supposed to occur at the entrance to the preserve at 8:30 AM, as others had before it, in order to provide a stark contrast between protected wilderness and the polluted landscape in the distance. But the young aide to the congresswomen directed us to the dirt road along the side of the landfill so that photographs would include "the steaming pile of garbage" in the background. "Actually, that's compost," I interjected. "Whatever," she said, and we moved. Along the road, the landfill was clearly visible and the compost pungent, which gave the participants something to discuss and recollect afterward with grins and grimaces.

When signs were distributed and people stopped arriving, the dozen or so protesters stood in one group and the two members of the press, there to document the event, stood in another. The structure of the rally was organized to feature Paterno, who read from a sheet of paper into a tape recorder at the front of the group. "This is about you," she began, "I am here at Four Corners landfill surrounded by four generations of a local family." The family in question was that of Maude and Gwen, who were there with some of their children, a grandchild, and their parents. The symbolism of the rally was straightforward. It was addressed to a broad public, sympathetic to opponents of waste importation. The demonstrators were meant to stand in for the local community as victims of the landfill, and Maude's extended family was intended as a synecdoche for all the families of Brandes, of the voting district, and of Michigan as a whole. This element of performance became clear to Bill later, when he asked Zimmler why he and his friends were not contacted by her office to attend another local press conference she had organized. She replied that press conferences were mere formalities for reporters and that it was not always necessary to have actual people present. This clearly bothered Bill, who grumbled that they might as well put up a backdrop of the dump in a room in Lansing and record it there.

Zimmler's speech also referred to the people gathered there, calling their struggle a "quality-of-life issue" that was being distorted by the political games of House Republicans and corporate greed. Like Paterno, Zimmler also supported the moratorium on landfill expansions, adding that "the financial playing field" also had to change. By this she meant that the cost of tipping fees on trash imports should be raised to further discourage the importation of out-of-state waste, a controversial measure opposed by the Republican majority. Jacob and Roy came forward next. They had prepared statements, which they read into the tape recorder held out in front of them, stating their grievances on behalf of the other demonstrators present and the larger community. Roy spoke about the terrible traffic problems and the "atrocious odor," and Jacob followed up as a local businessman who could not rent the apartments he owned along the county road because of the landfill. Zimmler followed Jacob's plea for change with a spontaneous remark on behalf of a wider public: "Our residents deserve to have parks to go to, deserve to hunt and fish in a clean environment!" This inspired nods of agreement and vigorous clapping from the crowd, which encouraged Paterno to holler a rallying cry: "I want action! Do you want action!" to which the crowd responded collectively, "Yeah!"

After the rally was over and some words were exchanged between the Brandes residents and the politicians, I was invited to join the former group at Lions Service for some cheesecake that Jerry had prepared. Back at the store, as everyone enjoyed dessert in the coffee klatch, some of their misgivings about consorting with politicians were voiced. Bill was particularly vocal about his skepticism: as he ate cheesecake that he knew would aggravate his acid reflux, he argued that the whole press conference was "done to look good, for the reelection" and not for the residents in general. Many were uncomfortable with the partisan nature of the event, which they felt was counterproductive. There was general agreement that something had to be done, but Bill and Jerry argued that all of them were helpless against the power of landfill capital. They insisted that in a few years the landfill would encompass the wetland as well, despite my suggestion to the contrary. Few really believed they could actually stop the landfill from expanding, even with the help of well-meaning politicians. This combination of skepticism and desperation did not change as they participated in more actions and meetings in the weeks leading up the important hearing with the MDEQ.

Subjection and Death

The subject of death came up several times in discussions leading up to Four Corners' expansion hearing. The morning of the rally, Maude's parents (who were Gwen's in-laws) began to recount stories of the group they had formed to oppose the landfill over a decade ago, which succeeded in preventing Four Corners from being sited across the street from their house. Maude brought up the fact that she was pregnant then. Smiling, she recalled that her parents were planning to have her lie down in the middle of the county road to stop truck traffic. Everyone laughed at this, but the conversation quickly turned grim as her father went on to describe some of the mysterious deaths that occurred at that time. The landfill company, then owned by a Detroit businessman many people believed to be a gangster, had warned one of the protesters, a local politician from Brandes, not to interfere with their plans. According to Maude's father, some time later his wife was found dead, having fallen down a flight of stairs. Not long after that, the politician himself died of a heart attack while driving. Maude's parents and their friends strongly believed that these two incidents were meant as a warning from the landfill owners to stay out of their way.

Many of those present had heard this story before, so it was partly being retold for my benefit, but it is interesting that they found it worth retelling at that moment. In addition to depicting, in a more hopeful vein, the secret and corrupt power they had faced once before and defeated, discussion of remembered corpses and Maude's vulnerable-yet-powerful pregnant body reflected the troubling position of politicized victims vis-à-vis the public they try to address and the people they enlist to speak on their behalf. The staged rally near the landfill depicted Brandes's residents as worth listening to, granted them a public voice on the foreign waste trade, only insofar as they were victims of environmental harm. It is possible to achieve political influence in this way, as the case of the Timiskaming demonstrates, but it can lead to an ambivalent sense of power through powerlessness, or ressentiment.[14] Images of wounded bodies presented on the day of the rally were related to a sense of embittered subjection, the logical outcome of achieving political capital as a suffering subject. For this reason, too, opposition to the landfill among Brandes residents could take the form of (somewhat) exaggerated claims about the conspiratorial power of landfill owners and the harm they have caused in the area.

Bill had many ideas about how to give opposition to the landfill more political clout (neither of which he would eventually pursue). One he described as a "quality-of-life" strategy, following the term used at the rally, which was the same as Maude's proposition at our first meeting, discussed in chapter 1: to organize a research study that would survey the local area for health problems. On one occasion, as we discussed how he might frame the issue so as to attract the right sort of public attention, Maude interjected, "I think we all have to die." Bill smiled and began to explain what he had really wanted to say to a state senator whom he had recently met:

> I was gonna suggest, the other day we were out there in the rain, when Debbie Stabenow and her whole crony group came over there, I thought "You know what if you really want to help us one of you guys would swerve in front of one of those [Canadian garbage] trucks and wipe three or four of ya out. If you really want to take it for the team!"

This comment caused uproarious laughter, but it quickly ended when Maude mentioned a well-known local tragedy from a few years earlier, when a woman was nearly paralyzed on the county road (which I will discuss below). These exchanges represent the tensions involved in achieving political recognition through subjection. Bill and his friends could amuse themselves by imagining the deaths of ambiguous allies, but in doing so they highlighted the inequalities that shape the recognition and legitimation of risks.

There are risks to political engagement through the wounded attachment of ressentiment, however. One claim I heard many times from Bill, Maude, and others was that if politicians from Washington and Lansing had to live in Brandes, then problems with the landfill would be taken care of immediately. But this presupposes that all those who *do* live in proximity to Four Corners share identical political sentiments. By implication, all local residents are homogenized into a single community, as was done performatively during the rallies near the wetland preserve and in the political demonstration in Lansing two weeks later. But not all residents felt as the local demonstrators did; nor did all the regular customers at Lions Service. I spoke with some who lived along the same roads, breathed the same air, and witnessed the same circumstances yet believed that nothing should be done or trusted the landfill company to protect them from harm. Some, in fact, would have suffered were the garbage to stop coming.

Politicizing ~~Canadian~~ Waste

Waste hauled in from abroad is not inherently political; nor is it necessarily represented as foreign. To become a matter of shared concern, any issue must be successfully addressed to a public with shared interest in its resolution. Such a public can be understood as a community of the affected who come to perceive the issue as their own.[15] The point of the political rallies was to bring the ghostly lines dividing Michigan from Canada to life through the concrete suffering of ordinary locals. At the first landfill rally, Roy, Jacob, and the four generations of Maude's family had been staged as synecdochic representatives of the impact of the Canadian waste trade on all Michiganders.

Many different publics are implicated in the circulation of mass waste—not only the people directly harmed by its disposal, but those who depend on its subtraction and addition. At the April town hall meeting that Paterno organized at the Calvin senior center, Jacob was publicly attacked by other residents who lived along the county road. Some blamed him for placing large signs against foreign waste in his front yard. Though many in attendance were sympathetic to his cause, some argued that his activism was making it difficult for them to sell their properties. Jacob searched the room of the senior center for allies and found none. He was, in fact, one of the few original demonstrators who had remembered to attend the meeting. Later, he begrudgingly removed the signs. This lack of uniformity among purported victims of the landfill became even more evident at the MDEQ hearing in mid-April, which pitted opponents of the landfill from Brandes and elsewhere against supporters from Harrison; irreconcilable differences between residents foreclosed any attempt to frame the hearing as a referendum on Canadian waste.

The hearing was not the only way in which public perspectives were considered: many citizens and private businesses had written letters and emails to the MDEQ, the Department of Natural Resources, and state politicians. I was told by the person in charge of the MDEQ's evaluation of Four Corners' proposal that the extensive documentation of verified and unverified complaints against the landfill played a large role in their assessment and shaped some of the terms of the application's evaluation. As discussed in chapters 1 and 2, inspections and regulations have repercussions for the day-to-day business of landfills, including the landfill tipping fees charged to customers and the larger market in waste contracts. Each phoned-in complaint, news story, or organized demonstration helps inform state regulation in some sense. The waste industry includes such variables in their decisions about where to site landfills and in projected costs of operation. "Host fees," for example, represent how the potential impact of local residents on the waste market is transformed from an unpredictable externality to a calculable cost of doing business.

The hearing was conducted by the MDEQ in accordance with state law, which mandates that significant proposals to change the conditions of landfill permits must go through a period of public deliberation in order to provide the agency with additional information for the construction permit process. At public hearings, laypeople are provided with a more open forum for dialogue with regulatory decision makers and waste-industry professionals. Settings like these are public forums or hybrid sites of debate and collaboration.[16] Party-organized press conferences and even friendly debates at Lions Service are, arguably, alternative public venues that aim to transform the market in waste services and may succeed in doing so. But for any set of concerns to dominate the qualification of waste services depends on the strength of the alliances that different actors actively cultivate through these public forums.

The MDEQ hearing included approximately one hundred people from Brandes and Harrison, along with various environmentalists, politicians, and landfill and state employees from throughout southeastern Michigan. Most of the people were from Brandes, and most, but not all, were opposed to the landfill expansion. However, those who spoke publicly were evenly divided for and against. The meeting, a little over two hours in all, consisted of an overview of the proposed expansion and the application process by Corey (the environmental engineer from Four Corners) and the MDEQ representative. Both were dressed neatly and were prepared with handouts and displays. They sat near one another at tables situated at the front of the room, facing the diverse audience. In part, their introductions were meant to ease public unrest and clarify common misconceptions about the issue. For example, Corey made clear that the changes would *not* expand the landfill past its existing boundary into hundreds of acres of additional wetlands to the north, as had been the local rumor. After the presentations, dozens of people waited patiently to make their comments, though not everyone at the public forum was given a chance, or equal time, to speak. The three dozen or so who did speak before those assembled included business associates of the landfill, politicians from Harrison and Brandes, representatives from Silent Pines subdivision, an environmental activist from the Ecology Center of Ann Arbor, and Congresswoman Paterno.

The comments ranged from passionate diatribes against the landfill or the MDEQ to carefully delivered statements and short, straightforward questions. Those who defended the decision to expand tended to be business owners who benefited from the landfill (such as a local gasoline supplier) or high-ranking members of Harrison's township government. Some of their comments were met with boos from the audience, which reflected not just differences of political opinion but a battle for control of the forum. One of the people booed most strongly was a Harrison resident who demanded to know why the hearing was not taking place in her township, the actual site of what she called "the dump." The resulting boos demonstrated, on one hand, the fact that many of those in attendance resented Harrison for reaping the benefits of Four Corners while others incurred the costs. At the same time, the boos indicated that a struggle was underway concerning which of those assembled should inform the MDEQ in their decision.

To which public or publics did this forum really belong? Brandes was selected as the site of the hearing, the representative from the MDEQ argued, because they asked to host it. Part of the agency's decision to grant the township's request was their conflation of local interest in the expansion with registered complaints, the vast majority of which had come from Brandes. What so bothered the Harrison residents in attendance was that the people of Brandes were being granted a privileged voice. But, as Harrison's supervisor informed the audience, and as he had told Paterno on many occasions, his township did not want the landfill either but was now dependent on it for 65 percent of its operating budget. They too were victims: a poor, rural town addicted to landfill capital and now threatened by efforts to cut back or eliminate a business that was nevertheless sustaining them. Some of those opposed to the expansion and to Four Corners' out-of-state waste contracts took offense at what they described as Harrison's position and defense of a source of revenue that threatened others' quality of life. In response, another Harrison resident attacked new residents of the large subdivisions in Brandes who had moved in after the landfill was built and now complained about the odor problems. Another woman from Harrison spoke out against the leaders of her township who would dare threaten her family with the landfill, revealing at the end of her diatribe that she planned to run against Harrison's supervisor in the next election.

These exchanges demonstrated a contest over who was the real victim, who the villain, and which locals were truly representative. They not only disrupted any attempt to enlist the community as belonging to one group or one set of interests; they also made problematic anyone's attempt to be a representative spokesperson for an entire street, township, county, or state. This was the difficult responsibility of Representative Paterno, one of the first to speak at the meeting. She began by thanking the MDEQ for their preparation and presentation, spoke briefly about the precise terms of the expansion and the landfill's postclosure agreement, and then promptly sat down. Bill, Jacob, Roy, and others who had followed her from the wetland preserve to Lansing in recent weeks were absolutely stunned. Where was the strong language of opposition to foreign waste? Why was she not acting as the spokesperson for their plight as she had at recent political gatherings?

From the very beginning, the demonstrators affiliated with Lions Service felt stripped of a public voice at the hearing. They were told that their anti–Canadian waste signs had to be left outside, even though, as they pointed out, the landfill was allowed their own props and signs supporting expansion, which made no mention of Canada. This perceived inequality structured the entire meeting from their standpoint. Maude felt that the people from the MDEQ and the landfill ought to respond to people's concerns and accusations more directly. As an information-gathering forum and not a question-and-answer session, the purpose of the hearing was to let people speak so that their opinions on the matter could be recorded and reviewed. But this made some of those present feel as if they were not being heard or properly recognized. Some felt that the MDEQ and Four Corners were acting as one, situated together at the front of the room, separate from those waiting to voice their questions and comments. In fact, I later learned that Corey had planned this—he wanted people to symbolically associate Four Corners with the legitimacy of the MDEQ and situated himself on stage accordingly.

While opponents of Canadian waste would see that hearing as a failure for a number of reasons, the perceived betrayal by Paterno would be the most memorable occurrence that night. In many ways, her actions encapsulated their sense of complete disenfranchisement by the state regulatory agencies and their apparent complicity with the landfill company. Later, Maude, Bill, Jerry, and others told me that "You could tell it was a done deal." This interpretation was associated with more exaggerated and conspiratorial claims:

that the public forum was merely a formality and that everyone—from the MDEQ specialist present to Paterno herself—had been "paid off." In some ways, this sense of mistrust, bordering on paranoia, was under the surface during every encounter with Paterno and the MDP. The local activists knew that they were being used, just as they were using others to reach a wide public. But their alliance was based on the assumption that Canadian waste gave Michiganders common purpose. For her part, Paterno was also disappointed in the hearing. Her aide later told me that she was hoping for five hundred or more attendees, and for more to speak out against the landfill and the expansion than did. Paterno's decision to remain polite and conciliatory can be understood, in part, as an effort to carefully manage her position as representative for both Harrison *and* Brandes residents. In a town hall meeting in Harrison earlier that year, Paterno had been verbally attacked by politicians and residents for suggesting that the Canadian waste trade should be stopped, without offering practical ideas for how the township would avoid going bankrupt as a result.

Paterno's actions were taken by the Canadian waste activists as a complete disavowal of the earlier vision of a unified community of protest that she had helped stage near the wetlands and elsewhere. In Bill's words, "She stabbed us right in the back." Days after the hearing, Jacob came across Paterno at a Big Boy's restaurant near Brandes and "[let] her have it," as Bill put it, for disappointing them. Jacob did not tell me about this confrontation himself, perhaps because he was embarrassed. But he did tell me that his experience at the hearing and the protest in Lansing had dissuaded him from engaging in politics. He felt used by their allies, whom he could no longer trust: "They say one thing to your face," he told me, "and then they laugh at you as you walk away. . . . I want to forget all of it. I hate to say it, but I've given up." Bill expressed similar concerns to Zimmler in a private conversation:

> It's great that you guys come out here beatin' your chest and say all these bills you introduce but the Republicans keep stonewallin' me . . . and guess what? You're right! So you can introduce legislation till you're blue in the face and all it is is like kissin' your sister cause it ain't goin' through, but it looks good on a press release: 'Look what I'm trying to do for my constituency and these Republicans won't help me.'

Frustrated with the failure of bills and rallies to effect real change, Bill and his friends stopped demonstrating and stopped trusting their representatives to help them. The public forum had forced them to recognize competing claims about the landfill. Moreover, they were confronted with the fraught attempts of political representatives and regulatory bodies to mediate these claims.[17]

The landfill's expansion became a political event, as the landfill's local opponents had hoped, but wounded attachments to Canadian waste did not dominate this process of politicization. With their signs and demonstrations, the Brandes activists had previously endeavored to stand in for (and marshal support from) Michigan as a whole, but the structure of the hearing upended this synecdochic performance. In symbolic and practical opposition to state regulators and politicians, local activists could only represent the interests of specific people and places, among others. As a consequence, the waste also lost its prior social identification as "Canadian," and what had seemed such a concrete manifestation of transnational pollution reverted to ghostliness. As the elusive object of so many competing claims, mass waste amounted to an anonymous source of profit/harm, just as it was for the landfill company and its clients: an abstract negative value demanding more air space.

This was not inevitable, however. It was a successful synecdochic performance, on the part of the Timiskaming First Nation and their allies, which had propelled waste to cross the border in the first place. Unmarked Brandes activists could not represent victimization from environmental injustice in the same way as could Timiskaming, who were able to draw upon a well-established racial economy that defined indigenous and white Canadians through mutual opposition. But in their desperation, they could (and did) draw on racial economy in other ways.

RACIALIZATION AND MASS WASTE

Canadian policymakers and journalists also commented on the waste trade with Michigan. Defenders of the practice tended to point out that Michigan exports thousands of tons of hazardous waste into Ontario annually, making the importation of several million tons of nonhazardous Canadian waste seem favorable by comparison. Bill and his fellow protesters were generally unaware and uninterested in this argument, not only because of their investment in local dilemmas, but also because they tended to believe that there was more for Americans to fear from Canadian waste and waste haulers than vice versa.

Twentieth-century U.S.–Canadian relations are often characterized as an unprecedented example of binational cooperation. Before Canada and the

United States were parties to NAFTA and the Kyoto Protocol in the 1990s, they had already been making transborder agreements for nearly a century. These included the International Joint Commission (1909), the NORAD missile defense system (1958), and the Auto Pact (1965), the last of which developed the Canadian auto industry, aided American corporate expansion, and ushered in the most profitable international economic relationship in the world. In 1972, the United States and Canada signed the first international agreement to restore and protect a transborder ecosystem—the Great Lakes. This eventually led to acid rain programs in both countries aimed at reducing sulfur dioxide emissions. This was later codified in the Canada–United States Air Quality Agreement (1991) and used as a model for the cap-and-trade policy framework incorporated into the Kyoto Protocol (1997) of the United Nations Framework Convention on Climate Change.

Despite real and unresolved conflicts between the neighboring nations, binational cooperation has been the prevailing narrative of the contemporary U.S.–Canadian border. And yet Canadian border-crossing into the United States is not a new concern. Over the past century, the northern border has been overshadowed by a national focus on migration from and through Mexico, but both have long been responding to and reshaping racial economies in the borderlands.

After the American Revolution, a new ghostly line divided the Old Northwest at Detroit, made concrete through waterways and associated bridges. This new international border promoted ideas of foreign and indigenous otherness. Throughout the 1800s, American officials worried that different Native American groups would migrate north, into territory claimed by the rival British Empire, and carry out raids against newly settled whites. By the end of the nineteenth century, amid rising concern about unchecked immigration from Asia and southern and eastern Europe, there were calls for more comprehensive protections against cross-border immigration from Canada. French Canadians, in particular, were racially marked by some U.S. labor groups as inferior whites or "the Chinese of the East." For the most part, however, immigrants from Canada were spared the racist nativism directed at other international immigrants, even when their numbers were actively reduced in the 1920s. The greater fear, then as now, was that unwanted foreigners would use Canada as a point of entry to the United States.[18]

The movement of mobile bodies across borders concretizes the ghostly lines of race and nation. My coworkers at Four Corners, as well as the Brandes activists, were especially interested in the waste coming from Canada, but they focused additional attention on marked bodies of Canadian waste haulers of foreign backgrounds. This attempt to relate material and social differences was manifested in the shared, but mistaken, impression that many of the Canadian waste haulers were "Arab." Some drivers were originally from Punjab or Trinidad, and there were no more than a dozen of them working at any given time. Waste circulation was thus racialized, regardless of what my informants felt about the waste trade with Canada, in an effort to represent transnational and occasionally traumatic encounters in familiar terms. As I will discuss in the next section, nationalist typologies of the foreign and the familiar take on life in everyday communicative practices and in the specific interests and understandings that inform them.

Racialization in Practice

Any racial economy relies upon specific acts of racialization, which help stabilize racial types and divisions. All of Four Corners' employees were born in the United States and self-identified as white when I worked there, and the interpretations they formed of social difference often reaffirmed white privilege. And yet "whiteness" is an unstable category. As discussed in chapter 4, the marked cultural figures of "wild men" and "hillbillies" were important to the history of migratory settlement and colonial governance in Michigan.[19]

Laborers at Four Corners would often acknowledge the fleshly semiotics of whiteness and class. Sometimes they would playfully racialize themselves even further. One laborer described our sunburns as "nigger neck." A play on "redneck," this was in keeping with the tendency—among laborers, at least—to call themselves "the landfill's niggers," "brothers," or "negroes." Given the prevalence of casual racism at the landfill, when employed as self-descriptions these playful and painful words acted as more than metaphor. Eddy once told us gleefully that he came home so suntanned one night that his girlfriend initially mistook him for a "nigger" and was afraid. Mistaken impressions such as these demonstrate the polyvalence of flesh in making and mocking social distinctions. At the same time, they reaffirm white privilege by allowing unmarked men to appear or label themselves as nonwhite, while disallowing the reverse. This structural asymmetry built into the semiotics of race is why this self-mockery was amusing for some of my coworkers. Even playful acts of racialization are constitutive of (even as they are constituted by) the entrenched racial economies of settler societies, in general, and the specific interpretive repertoires of the greater Detroit area, in particular.

Unlike playful acts of self-racialization, racial epithets were used only to describe nonwhites behind their backs. Most of my coworkers also labeled Punjabi and Trinidadian truck drivers as Arab (pronounced "Ay-rab") and sometimes as the more derogatory "towel head." On rare occasions, some of my coworkers also used the term "Pakis" or "Hindus," most likely learned from relatively unmarked Canadians who used it more frequently. The distinct cultural figure of the Arab immigrant is part of an interpretive repertoire of social types deployed in southeastern Michigan and throughout North America. The ability to depict social difference in this way comes from a place of white privilege, which serves as the unmarked background against which Arabs may appear to be polluting and threatening. This is a common racializing ploy in North America, whereby generic whites become the default kind of person, against which opposed racial categories are marked as different. At Four Corners, truck drivers were typically referred to as "Arabic" when they deviated from an unmarked "white" standard and the speaker wished to single them out. Like "Arab," racially marked epithets were employed most often in those circumstances in which Sikh, African-American, or otherwise foreign individuals figured prominently in workplace narratives. Identifying people as "Arab" stabilizes the otherwise ghostly lines of racial and national difference, which are elusive because of both their imaginary basis and the relative limitations of the interpretive repertoires that may be available to speakers and their audiences. The misidentification of some Canadian drivers as Arab could be described as a form of predictive racialization or racial abduction. Misrepresented as Arab, the possible actions and motivations of transnational waste haulers were rendered interpretable. For my coworkers and local activists, this meant that Arab waste haulers came to be seen as sources of social and material pollution, responsible for an array of social ills—from poor English skills and bad driving to drug trafficking and terrorism.

White Privilege and (Mis)identification

One afternoon, a Canadian truck driver named Hasan approached the snack shack and insisted on buying me a coffee, with what sounded like a South Asian accent. I agreed, and we exchanged only a few words before Tanya jumped into the conversation, asking with a friendly grin, "*Where* are you from?" "Trinidad," Hasan replied. "India?" Tanya responded, substituting her own answer for one that didn't register. I stepped in: "No, Trinidad," and (as the self-appointed broker of cultural difference) tried to explain the distinction before I was cut off by an impatient Tanya—"I was never good at geography in school; I'm only interested in the future." Hasan went on to explain how the Trinidad he'd grown up with was no longer a part of his future because it had changed so much and he had no desire to visit again. When he walked away, leaving me with my free drink, Tanya informed me that he was a "nice guy" and a "good customer." Once he was out of sight, she remarked, laughing, "Do you think his wife has a jewel in her belly button?"

For Tanya, as for many unmarked North Americans, foreign identities tied to unfamiliar places seemed caught up in an exoticized past, as opposed to the future. But Tanya's refusal to hear an explanation of Hasan's background can also be seen as an attempt to reach out to him as a possible recurring customer, as if to say that for her, his markedness is of no account. While this was likely intended to be a friendly gesture, it misjudged the asymmetrical nature of racialization. In North America, white privilege means receiving the material, psychological, and symbolic benefits of racial hierarchy without having to be aware of it. On one hand, whiteness is the absence of being marked as other: the ancestry or social background of someone white is, by definition, of no account in the United States, which is associated with the ideal of modern self-creation and the American fantasy of starting over with no past. On the other hand (and for this very reason), whiteness can also mean having the freedom to not know others—to ignorantly misidentify—at no cost.[20]

Tanya playfully referred to two of her repeat Sikh customers as "the double doubles"—since they always appeared together and routinely ordered two sugars and two milks for their tea, which they drank boiling hot. Like most customers, Tanya's double-doubles tended to appear around the same time every day and, though they bought only tea, she tried to make them feel welcome, despite significant barriers to mutual comprehension. On one occasion, I witnessed the three of them sharing a laugh when she pretended to put too much sugar in their tea. And yet Tanya routinely misidentified them to me as "Arabic" and "Muslim." While Tanya actively tried to cultivate a diverse clientele, her predecessor at the snack shack (and her sister-in-law), Donna, once told me that she had trouble being friendly with the Sikh drivers "because of everything going on in Iraq," referring to the occupation, which was then just over a year old. In both cases, being unmarked means being free to act on misidentification without consequence.

Sikh drivers were not always misidentified by visitors to the landfill, however. One afternoon at work, I was alarmed to notice a black SUV parked near the top of the southern slope, looking down on where I was cutting willow trees on the side of the road. My initial thought was that they were watching me and judging my work performance, trained as I was to treat those in vehicles perched above me with suspicion. Later, I learned that it was a class of architecture students from the University of Toronto whose professor was doing work for the landfill and taking his pupils on tours to different waste sites in the area. As I was leaving for the day, I saw the SUV parked outside the landfill, so I pulled my car over to introduce myself. I was a bit self-conscious, reeking of sweat and filth from a hard day of work, but the students were excited to talk about the places and things they'd seen that day. They encouraged me to see a dump farther north where they'd been that morning, which one characterized as "sublime," and were interested to hear about the results of my research so far. Their professor was white, young, and spoke to me only once. I was explaining relationships between landfill workers and the people they took for Arabic truck drivers and was abruptly interrupted: "But they're Sikh!" he said with a wide grin, prompting his students to laugh. His confident interpretation was based on the turbans that he'd seen some of the drivers wearing and their long, unshaven beards, signs of an ethnic type that both of us had learned at some point was stereotypical of Sikh. Indeed, this is precisely why those adorning themselves in this way chose to do so: to be seen by others and recognized as such. Ironically, as a result of the historical formation of their diaspora, male Sikh bodies are meant to be on display and fetishized through the adornment of the "Five K's" (see below).[21]

I was uncomfortable with the visitor's knowing laughter and casual dismissal of my coworker's ignorance and racism. Partly, this was out of awareness of the limitations of the latter's interpretive repertoires, for which they were not entirely responsible. The privileged male academics knew better than the women who ran the snack shack, yet we were all participating in a game whereby unmarked whites get to (mis)identify marked others without needing to worry about being identified in turn. But it also made me uncomfortable because, were this standard of typologization to be rigorously applied, those Sikh men without turban and beard could be misidentified as essentially *not-Sikh*.

The Sikh driver I came to know best was a thirty-something man named Bula, who spoke better English than most of the other recent immigrants to Canada I met and was incredibly friendly and outgoing. We conversed when I was stationed at the "hotbox," where trucks with frozen loads were sent in the winter to thaw. The hotbox was a simple operation: one landfill employee would guide the trucks into the large building, close the doors, and position and ignite four or five powerful propane heaters that spewed flames underneath the trailers for ten or fifteen minutes at a time. On very cold days, drivers would park in a line of ten to twelve trucks, and some would come to talk to me as they waited their turn. Bula and I would talk as he waited in line or in the hotbox. He had an easy-going demeanor and would follow me around as I tinkered with the heaters or paced outside in the freezing weather to cool down from the overwhelming heat of the box.

Bula was clean-shaven and wore no turban, which led me to wonder if he was attempting to avoid the exoticization and discrimination that some of his friends and fellow countrymen encountered while crossing the border. One day, waiting for a friend of his to finish dumping up top, Bula and I spoke about being Sikh. "I am not a religious man," he said. "It is too hard here." He told me it was difficult to eat vegetarian in Toronto, or to know whether dishes and types of food were kept separate during preparation. He explained to me the principle of the Five K's, which are the cumulative garments and bodily habits assumed by religious men, including some of his coworkers: uncut hair, a kirpan sword, an iron bracelet, a wooden comb, and a special undergarment. Bula hoped to become a true Sikh one day, an *amrit-dhari* committed to the Five K's. He hoped to do this later in life, when it would not be so complicated. He also confirmed what others said, that it was more difficult at the border for what he called "nonwhites" because of their appearances and accents. He claimed not to be bothered by this: "It is reasonable," he would often repeat.[22] Despite Bula's pragmatism, the truth is that he had no choice whether or not to deal with the consequences of (mis)identification; it was something he lived with every time he crossed the border or entered Four Corners.

As I discussed in chapter 2, landfills are not arranged to promote worker solidarity; there is a significant division of labor between the more skilled machine operators and mechanics and the slightly lower-paid, easily replaceable truck drivers. Not only do the latter work for different companies, but their jobs are more likely to put them at odds with landfill employees in the context of the workplace. Most truck drivers are motivated to come and go as quickly as possible in order to meet their deadlines, particularly if they are working in tandem with another driver who needs the truck back in Ontario by a specific time; and even those with time to spare do not wish to linger up top, where they may get stuck or pop a tire. Landfill employees, on the other hand, are more interested in maintaining orderliness, avoiding accidents, and keeping a low profile while on site. Needless to say, this results in occasional conflicts, with the landfill employees frequently complaining about "bad driving" and drivers accusing some operators and managers of being "bossy" or slow to get them down off the hill.

The use of "Arab truck drivers" as a popular trope in workplace discourse is not exclusive to landfill workers. I have also heard American and Canadian truck drivers criticize the Arabic drivers or "turbans." In general, this was done in the context of expressing frustration with drivers who seemed poorly trained or unfamiliar with the elaborate routines and signs used to maintain safety and orderliness up top around the exposed dumping face. In popular gossip about the Arab truck drivers, poor English skills were often mentioned alongside bad driving as the most significant barriers to getting the job done. Many of the unmarked Canadian drivers and landfill employees shared a belief in the importance of plain speech and direct, referential meaning when exchanging signs. The tempo of work on the roads and up top was structured by hand signs and relatively brief exchanges over the CB radio. Machine operators used the CB to impart English directives to truck drivers when necessary in order to maintain a degree of orderliness and facilitate efficient and fast-paced disposal. Orderly talk and an orderly workplace are ideologically interwoven.

Over the landfill's CB channel, the tendency was for short bursts of communication, usually referencing events and activities that both participants could visualize from their vehicles: "You're alright. You can get by, c'mon truck!" The number of participants was limited only by geographic range (sometimes people communicated from the highway that they were almost there), but often this consisted of short exchanges between two or three people in the same vicinity, with a brief statement followed by a prompt reply:

"Hey, where we going with this?"
"Put the garbage with the garbage, ten four."

"Mike, Dave's comin' around to hold that up for ya."
"Alright."

"Hey Tanya what's your special today?"
"Barbecue chicken on the grill or hamburger, macaroni salad baked beans."

As a linguistic register, CB talk at Four Corners privileges direct, literal interpretations of speech acts.[23] At the same time, as people become regulars and grow more comfortable with the landfill, they may extend CB talk beyond its typical propositional framing:

> [Singing] "Lonely Days! Lonely Nights! Where would I be without my woman!"
> "Hey don't give up your truck-driving job!"

Even in such instances, however, talk remained close to the stereotypical form: short, dyadic exchanges confirming adherence to the underlying speech register.

Like bodily adornment, CB talk was also involved in positioning speakers in relation to one another as types of persons.[24] Arab or Canadian speakers on the CB were potentially marked, through their participation in these speech interactions, as certain kinds of personae set apart from the unmarked, American workers. One obvious violation of CB interaction was to use language that could not be understood by everyone, or to be someone who didn't understand the dominant language of the site. Sometimes, drivers of different backgrounds might be able to understand fragments of English but unable to form comprehensible and acceptable replies; they would later be characterized by landfill workers as being unable to speak English. Occasionally, truck drivers with difficulty comprehending or speaking this register of English would rely on a bilingual assistant who used the CB to translate English orders into Punjabi, for example. Although this allowed the landfill to operate relatively free from complication, it raised the ire of some workers, who sometimes said over the CB, in a loud and irritable tone, "Speak English!"—though, for the most part, speech interactions over the CB and in person were relatively polite. It could be argued that dependence on a third person, acting as translator, adds another layer between operator and driver in what is supposed to be direct, unmediated talk, thereby threatening the operator's authority over the dissemination of signs and, hence, the ordering of relations up top. The assumption seems to be that failure to use the CB in ways that allow for simple directives and responses that landfill workers can understand disturbs the labor process, aligning failure to get the job done with failure to learn English.

There are other ways in which language difficulties and social types are indexically linked to the interruption of work flow. One American truck driver, working out of a company in Lakeside, argued that dependence on bilingual drivers leads to bad driving on site and on the highway. "Those

Arabs drive in herds," he told me, "so that they don't lose sight of their leader, the one who speaks the most English." This leads them to speed, drive too slowly, or pull over suddenly in order to keep pace with one another, he claimed. The assumption that proficient bilingualism confers a leadership status on Sikh drivers is debatable; it is certainly true that those who are not competent in speaking English are somewhat dependent on those who are. Not surprisingly, Sikh drivers at Four Corners tended to associate fairly closely together. However, apparent herding was quite common among many of the truck drivers I observed coming and going: they stayed in proximity to one another to chat on the CB during long hauls, meet together for lunch or coffee along the way, and give each other assistance in case of emergency. In fact, the man who informed me of the connection between herding behavior and linguistic incompetence was actually waiting for a friend to help him because his vehicle had broken down! Even where relatively marked drivers resemble their work associates, in others words, the tendency to label them as representatives of an unfamiliar social type—that of the immigrant or the Arab—may lead to the misrecognition of similarities between them.

I have tried not to overemphasize the suffering of misidentified truck drivers or romanticize the routine politeness of (most) transnational encounters at Four Corners. Bula, for one, thought he had a good job. His biggest concern lay with his children, whom he wanted to teach to speak Punjabi, and his father, whom he wanted to support but who was increasingly restless being stuck at home all day and wished to return to Punjab. That these were Bula's most compelling concerns is arguably in keeping with the historical formation of the Sikh diaspora and the predicaments it creates in terms of Sikh identification. His father's restlessness and his own desire to care for his family in the right way expressed some of the core aspects of the Sikh subject of diaspora. Those Sikh drivers who adopt some of the Five K's as they cross international borders also perform the movement of the Sikh body as a global sign in circulation—one recognizable to other Sikh as well as to intellectual elites like myself and that visiting class from Toronto. For less visible and representative Sikh, like Bula, not living up to this standard may feel like a failure, for the moment, to be a proper Sikh. Additional public surveillance and alienation may not seem bothersome by comparison, although it will depend on how specific racializing acts and the broader racial economy concretize the ghostly lines of race and nation in each case. A Sikh friend of Bula's pointed out to me that being "too religious" could get immigrants into trouble, adding that the hauling company never hired Muslims because "they have names like Ali or Mohammed." If this were true, then the presence of Sikhs at the landfill and their subsequent misidentification as Arab were made possible partly by the absence of people with Muslim-sounding names (who were possibly misidentified as being religious). Forms of racialization are imbricated, and so too are their compounding social effects.

Terror as Usual

Perceptions of the U.S.–Canadian border changed significantly after the terrorist attacks of September 2001. For a time, it was wrongly believed by prominent U.S. politicians that Al Qaeda hijackers had entered the United States from the north to carry out the terrorist attacks of 2001. While untrue, this mistaken belief was a symptom of growing border anxiety and securitization. On the day of the attacks, border traffic was brought to an immediate stop, disrupting the busiest international passageways in the world. Canadian truck drivers told me that the wait time to enter southeastern Michigan was more than six hours at the border, creating a thirty-mile traffic jam. For over a week, the Ambassador Bridge connecting Windsor and Detroit was closed. For months afterward, those crossing the border were subject to frequent searches, sometimes by military personnel. Certain areas became the focus of greater security and surveillance measures than others, as part of a general shift in border policy. For some pundits and politicians, the Canadian border raised unique concerns that *la frontera* with Mexico did not. They identified the largely undefended border and more liberal Canadian policies toward immigration as grave security threats. As it had been centuries before, Canada was identified as a source of possible social contamination.[25]

Key to this idea of pollution is the cultural figure of the Arab immigrant, which took on importance for some Michigan residents as their lives were being transformed through border violations, both material and symbolic. With some incidents of transnational encounter, the predicaments associated with the Canadian waste trade are more readily identified with the war on terror and its racializing nativism. The racialization of persons into a fixed type—their mass substantialization, if you will—here serves as a counterpart to the personalization of mass substance, or the transformation of waste from anonymous and abstract negative value to a vital concern.

Some Brandes activists actively sought to support their cause by appealing to the climate of suspicion and surveillance surrounding Arab and Muslim people in the years following 2001. The first time I had heard this was the morning of their very first demonstration: "This might make me sound like a racist," Bill began in his characteristic style, "but ninety percent of them Canadian truck drivers are Arabic. Shouldn't that be a homeland security issue?" That morning, I had made the mistake of trying to correct him. The first time I corrected this misidentification, I was politely dismissed: "Six of one, half dozen of the other," Bill said. Another protester agreed: "They're probably about as different from each other as Brandes is from Eatonville." Before I could respond, someone tactfully changed the subject. Months later, having spent more time with the group of activists and their friends and families, I realized my error. Arab misidentification was not about hapless ignorance of geography or anthropology. Rather, it was a selective interpretation, in two senses: first, as among my coworkers at Four Corners, "Arab" had been chosen from the limited interpretive repertoire of meaningful social types available; second, it was applied without any effort to test these interpretive criteria (for example, by engaging with a driver about his origins).

At the initial rally at the wetland preserve, Bill approached Paterno with this idea. I then realized how carefully his idea was crafted to match popular paranoia. In the spring of 2006, the Dubai Ports World scandal was receiving a great deal of attention in the national media. The Bush administration was being attacked for allowing a company owned by the United Arab Emirates to take over control of six large U.S. ports, a deal disparaged as a threat to U.S. security. At the rally, Bill tailored his statement accordingly: "You know that it's Arabs that own all trucks that haul garbage out here. Now that's a homeland security issue!" This was also untrue. The different hauling companies bringing waste from Ontario to Four Corners were all under Canadian ownership, though a Dutch company had recently negotiated the purchase of the company responsible for hauling Toronto's garbage. Paterno carefully skirted the issues Bill was attempting to raise, recognizing some of their unsavory implications perhaps, but his strategic racialization of Arabness was no different from what he had seen prominent politicians and pundits do for years.[26]

For some, racialization of waste haulers was not just strategic xenophobia but an effort to cope with painful trauma. I met Sam, a thirty-five-year-old laborer for a construction company, while he was exiting the woods along the north side of the county road leading to the landfill, carrying a basket of morel mushrooms he'd just picked from a wild patch deep in the forest. Sam had lived just down the road until moving three years before. He still returned periodically to hunt and pick wild mushrooms where an old house had burned down in the forest, providing nutrient-rich soil ideal for morels. Sam didn't like to come back very often, I soon discovered, because his previous life had ended in front of his old house, where a Canadian truck driver had struck his fiancée's car. The couple had met while working construction and had gradually become inseparable over the years: buying a house in Florida together, skydiving, and spending all of their free time and extra money meticulously assembling a miniature house with expensive replica furniture and tiny appliances. In late March 2003, on the day of the accident, Sam was playing with his fiancée's five-year-old son and waiting for the school bus to pick up her ten-year-old. In a hurry to get to the landfill, a waste hauler clipped the side of her small sports car as she sat idle, waiting to pull into their driveway. She was severely injured and nearly paralyzed. According to Sam: "They called her a broken girl. . . . She broke this arm, broke her femur right here at the ball, tore this part of her face off, broken her jaw in four places." But that wasn't the worst part of the accident, he told me: "I think the mental state turned out to be the worst. Nobody's ever the same after that, whole family's out of control, everybody's turning on everybody . . . and me being just the boyfriend, you know, I'm takin' the heat of it."

For the first few months after the accident, the boys had trouble with discipline and would try to turn their mother's family against Sam. The younger one began having unpredictable seizures. A big part of the problem was that "she wasn't the same person anymore," Sam said, but was prone to unexpectedly crying or lashing out. Their relationship couldn't recover from the accident. A big part of the problem was the daily truck traffic passing by their house: "Trucks are constantly goin' by here, but once you get hit by one right in front of your house and you gotta sit and listen to it . . . it's very stressful." Sam admitted this was a dark time for him, when he began to feel alienated from the woman he loved and her boys, whom he'd once begun to think of as his own. As he listened to the truck traffic and relived the incident over and over again, he was taken back to the "Arab" who had struck his wife's car and had smiled at him at the scene of the accident. That smile made Sam furious. "It was a grin," he told me. "I almost went for his throat."

Later, Sam learned that the same driver had had his license revoked when he'd struck another person. He began to suspect that the accident had been intentional: "I was thinking it was a conspiracy; I started thinking it was a part of terrorism, you know what I mean [nervous laughter] I was thinking it was a internal terrorism thing—they're just goin' round hittin' people, just zoom into people." After the accident but before he had moved out, Sam was taking time off from work and spending it at home. He started looking for proof of what he suspected, that the accident had been a terrorist attack. "I got to the point where I just sat there for hours and hours, videotaping trucks goin' by all day and night, taillight after taillight after taillight." He was looking for evidence against the landfill as well, of suspicious activity or broken rules to use in the event that they needed it for court. Mostly, Sam was looking for some pattern behind the endless line of trucks, sometimes five hundred in one day—at the busiest times, one passed by every five seconds.

Sam's assumption that the driver was Arab was about more than possible misidentification; it offered him a way of interpreting the unreal trauma that befell his family and that, he believed, ruined his life at the time. His short-lived obsession with videotaping the truck traffic, in order to capture some secret or discover a hidden plot, was an extreme version of the border anxieties that I have documented in this chapter. In their own ways, the residents of Brandes who criticized the waste trade were engaging in acts of surveillance and suspicion on behalf of a state that leaves the problem of waste importation to the invisible hand of a transnational market. Landfill workers tended to racialize Sikh drivers as Arab, to different ends, but continued to rely on a social poetics that linked unwanted matter with unwanted people. Taken to its limit, as in the case of Sam's personal tragedy and its aftermath, concern about what the truck drivers were hauling could mutate into paranoid fears about who they might really be.

Landfill workers and local residents rarely saw eye-to-eye, but both sides enjoyed gossiping about two purportedly Arab truck drivers who were arrested in possession of a few hundred thousand dollars as part of a drug-trafficking sting several years before. Brandes residents knew very little about this incident, but most were heavily invested in its symbolic import. According to rumor, Arab waste haulers had smuggled everything from illegal immigrants and drugs to dead bodies across the border, hiding them inside their trucks as they passed through.

Smuggling is not mere fantasy. It is, in fact, more difficult for sensors, scanners, and dogs to notice a bomb, a bag of drugs, or a suitcase of money when it is mixed in with so much rotting waste. The sheer volume and heterogeneity of transported waste leads one to conclude that waste haulers could contain almost anything. Here, a metaphor of foreign concealment and threat becomes metonymic. The practical reality that waste trucks could conceal almost anything combines, when it is transnational, with a sense that foreign companies, countries, or laborers may possess concealed motives. When Bill,

Sam, and others identified a Canadian as Arab, they questioned that person's hidden agendas with implicit reference to this pervasive atmosphere of fear and suspicion. The very alienness of the drivers identified them with the contaminating substances (both real and imagined) that they carried into the United States: they were unwanted, human waste.

IMMATERIAL PUBLICS

The conventional anthropological portrayal of purity and identity assumes that material classifications are derived from social ones, but the reverse is also possible. In this case, the way materials are separated and circulated, with ever more categories and ever farther destinations, results in fresh divisions between people (see Spyer 1998: 8). Modern waste management requires that some are exposed to the waste of many, thus always leading to disproportionate exposure to harm and stigma. To the extent that this circulation is implicated in the racial economies of North American settler societies, it not only apportions risk where one would expect but creates emergent divisions and tensions where one would not. It is as if the farther we want to distance ourselves from our transient waste, purging material difference in a quest for reconstructed sameness, the greater the social divisions that can result.

When Timiskaming First Nation activists successfully fought against environmental injustice in Ontario, they inadvertently contributed to the exportation of Toronto's waste to communities they likely knew nothing about in southeastern Michigan. Similarly, when Four Corners was forced to alter its sludge contracts, in response to local complaint and opposition, the waste still had to go somewhere. By 2007, Toronto officials knew they would no longer need to use Michigan landfills because they had come to an agreement to use one located in Oneida First Nation lands in southern Ontario. Brandes residents had no reason to be aware of their connection to the Oneida or Timiskaming, how their lives were entangled through the absence and presence of Toronto's waste. A racial economy continues to dictate waste distribution in North America, as has been well established by the environmental justice literature. To struggle against this broader reality means recognizing that the movement of waste is about more than the waste itself—it is shaped by, and helps maintain, longstanding social hierarchies, including the ghostly and fleshly lines of settler societies.

This environmental injustice is not inevitable; it is a logical outcome of uneven waste distribution. If people are separated from people, this makes their unequal separation from waste easier to manage. If new separations are required, at greater distances and costs, then fresh divisions between people and places may be the outcome. Urban Canadians around Toronto versus rural Michiganders on the outskirts of Detroit? Such an opposition was not part of an existing interpretive repertoire for either group and was therefore difficult to make sense of. It is for this reason that a well-established racializing repertoire provided some Michigan activists and residents with a means to reflect on their transnational encounters with social and material difference. Without attending to the politics of waste circulation, waste haulers may come to appear like their cargo, a social bad, something that ought to go somewhere else to be dealt with by someone else. If cast in this way, resistance to waste imports can foster nativist concerns and the reassertion of white privilege, rather than broader alliances with other, similarly marginalized waste recipients.

Shortly after my fieldwork concluded, concerns about a terrorist threat originating from Canada were followed by fears concerning the international spread of SARS and Avian Flu, not to mention spirited debates concerning Americans without health-care coverage crossing the northern border to purchase prescription drugs. Polluted and polluting bodies may become figures of transnational concern apart from waste circulation. The reverse is also true. The same Michigan activists and residents who used racism and nativism to interpret their predicaments have also attempted to alter their circumstances in many other ways. The possible risks and actual harms of living in proximity to a dumping site provided opponents of Four Corners with alternative forums in which to change the course of the North American waste trade.

The Brandes activists I came to know eventually agreed to participate in a successful class-action lawsuit against Four Corners, which was settled out of court. For the first time in over a decade, it appeared that the landfill company would have to directly acknowledge their criticisms. The financial settlement with the landfill company, like the social movement itself, did not take aim at the conditions that made transborder waste circulation possible, however. The suit focused on the quality of life and enjoyment of property that Four Corners had allegedly denied nearby residents through excessive odors. Once again, the residents' main recourse was to politicize their status as victims of specific harms rather than interrogate what made such victimization possible. The subtraction and disposal of mass waste means that some people will be disproportionately affected when it comes to rest. Despite Bill's insistence, there is no imaginary "Siberia" where waste can vanish without a trace. This common reliance on the uneven distribution of mass waste is what connects most North American "backyards." But landfills are designed to make clues of this imbalance vanish, to clean up the scene of the crime so completely that we don't suspect one has even been committed. All that remains is the productive absence of what was once there. When that redistributed waste comes from and goes to distant places, the challenge to recognize waste management as a social relationship is even greater.

All waste activism is in danger of becoming a species of NIMBYism (not-in-my-backyard-ism) for the simple reason that waste has a tendency to narrow one's focus. When the problem you are struggling against manifests itself as smelly sludge dumped in front of your house, wafting down your road, or hauled into your state, it is easier to miss out on the broader social and political realities involved. That being said, rallying to force waste somewhere else is not an insignificant change for those affected by its disposal. When nearby landfills close down or waste markets dry up, these people may now partake of the same freedom from transience once denied them. At first, Brandes had hoped that their efforts had stopped the sludge contracts with Detroit and Toronto for good. It is no secret that this concession was made as part of Four Corners' successful application for expansion. The MDEQ had been trying to eliminate their sludge contracts for several years, which they held responsible for noncompliance issues at Four Corners and other sites before them. Management at Four Corners had actually planned to stop taking in sludge once before, in the face of incredible odor problems in the spring and summer of 2004, but eventually elected to keep the contracts when the situation began to improve and a new odor-prevention plan was implemented. If not for the hundreds of odor complaints amassed over the course of several years, the MDEQ would have had less leverage in the negotiation of Four Corners' expansion. In this way, local residents succeeded, through their reflexive feedback to state and county regulators, in reshaping the market in waste services to their advantage.

During one of their first meetings prior to the settlement, in a basement in Eatonville, their lawyers took some of the credit for stopping the sludge contracts, which had been announced the day before. The reason that the landfill was now making concessions, they argued, was because of the financial threat that the class action represented. Four Corners hoped that people would give up on seeking damages if the main source of their concerns were to stop. Many people, including Bill and Maude, found this claim convincing. Bill mentioned at the meeting that no one should be confident that the landfill will keep sludge out of the community. As he said, "Look at their track record," adding that the community should maintain pressure on the landfill, bellowing "You can't let up!" as others nodded in agreement.

Bill's suspicions eventually proved correct. The landfill did take sludge again. Just as it is possible to feel empowered and not be, however, it is possible to feel disappointed and betrayed and still participate in changing one's circumstances to some degree. When the landfill resumed taking sludge, it was only the treated kind, which, at least according to some employees, made the loads completely odorless. Bob also used local complaints as justification for introducing a program to clean trucks before they departed the landfill, to prevent them from trailing garbage along the county road. Given the importance of odors and traffic to the people of Brandes, these were not insignificant changes. But there were other consequences of temporarily stopping the sludge that Bill and his friends did not imagine and would not have necessarily wanted. As promised, Bob fired three operators in the interim period, claiming that changes in sludge contracts were the reason.

If residents of Brandes accomplished some change by engaging in demonstrations, rallies, hearings, and lawsuits, they did not seem satisfied the last time I spoke with them. The enormity of what they had not chosen for their lives or their communities confronted them on a daily basis.[27] Yoking their ambitions to racialization or recognition as victims, they reaffirmed their dependence on the intervention of regulatory agencies, the media, politicians, and legal firms, not as equal collaborators but as wounded subjects to be protected and defended. This may have been a partially effective strategy, but they employed it only with ambivalence and uncertainty. Faced with a post-agrarian, post-industrial, and now post-housing-bubble world dominated by challenging new forms of global connection, they could count on little else.

Conclusions

Who keeps us safe from ourselves?

ROBIN NAGLE, *Picking Up*

WHETHER OR NOT THEY WORK and live within a landfill's orbit, North Americans tend to share many common concerns: they watch and worry as their communities change, try to stay clean and live longer, aspire to improve their lifestyles and raise successful children, shop for and buy things, and struggle to make sense of social and cultural difference in a world of transnational markets and migrations. Being routinely exposed to mass waste can amplify these familiar concerns. For example, staying clean becomes a serious challenge rather than the brief, quotidian practice it is for many others. The differing intensity of waste relations, homologously restaged through comparison, helps bring into relief how much North American life has become dependent on the productive absence of waste.

North American landfilling systematically exposes the few to the transient materials shed and forgotten by the many. By caring for our mass waste, waste workers indirectly care for us. I did not hear my coworkers put it this way, but many of them shared the sentiment that everyone's waste has to go somewhere and theirs was a necessary task. Other paid professionals are more often credited with caring for us. In chapters 1 and 2, I compared landfilling with medical sites and practitioners: landfills create homogeneous spaces free of filth and decay, much like those in hospitals; landfill workers surgically remove layers of skin from landfills to stuff more waste in; surgeons do dirty work without the accompanying stigma; and so forth. This was meant to encourage the reader to imagine waste workers as involved in projects of care for others. People rarely compare paper pickers to neurosurgeons, but this is arguably a consequence of just how well the former care for us. The successful subtraction of mass waste from (most of) our lives also separates waste producers from waste workers. The absence of sickness or injury might be credited to medical expertise, but we more readily take for granted the absences that waste subtraction provides (the fact that neurosurgeons depend on new and sterile instruments, for example).

Not all acts of caring are equally visible or are recognized as such, even by the carers themselves. One could say that all those who live near landfills also sacrifice for the rest of us. After all, their more direct exposure to all our waste occurs only because it has been combined and amassed in one location, rather than more equitably distributed. Host communities like Harrison are also paid for acting as custodians of other people's waste. Some wish this were not the case. Indeed, acknowledging what those who work and live with Four Corners do as a form of care does not mean that waste firms, their employees, and those who regulate them are beyond reproach. People in waste management make mistakes and could improve. Just because landfilling is done on our behalf does not mean it should stay the same.

There are at least two ways we might begin a project of reform. One is to see mass waste disposal itself as the problem. If this were our starting point, then it would seem logical and necessary to transform North American waste-management infrastructure. We might change our preferred methods of disposal and treatment, for example, or explore new ways to redirect and reduce waste streams. I will discuss these options below and argue that, while important, they are insufficient. If we recognize landfilling as a kind of care, we also have to reckon with why we refuse transience and not just the consequences of this refusal. Vital infrastructures like mass waste management can be changed only if the many problems they were designed to solve are also called into question. Some of the human problems landfilling addresses are universal. They may still conflict with one another, however, like the interconnected need to defecate and need for clean water. Many other problems are culturally and historically contingent, but no less taken for granted, like the choice to buy things new or renovate our homes.

In this concluding chapter, I will not provide an exhaustive evaluation of alternative waste practices or solutions to replace landfills, which is better left to sanitary engineers and policy experts. Instead, I examine more closely the representation of the landfill as a hateful problem and what this may conceal. In the course of doing so, I suggest a new image to associate with waste—that of a planetary core—and a new policy proposal concerning how we should relate to landfills—we should be buried in them. These are meant as provocations in order that we might better appreciate what stands in the way of imagining a future with(out) landfills.

ZERO WASTE?

I have argued that contemporary North American life cannot be disassociated from landfilling, but things could have been otherwise. One excellent counterexample is that of Sweden. In 2012, Avfall Sverige, Sweden's public–private waste-management consortium, announced that owing to the success of recycling programs, a mere 4 percent of household waste was being landfilled. In the spring of 2014, they claimed the figure was now less than 1 percent, making the utopian possibility of zero waste seem attainable. It should be noted that there are fewer people making waste in Northern Europe, and less space is available there for dumping. The historical rise of the Swedish middle classes also relied on waste subtraction and reconstructed sameness.[1] Yet they did not embrace the invisibility of mass waste as fully as their counterparts across the Atlantic did. One might attribute this apparent recycling gap to a broader distinction between Northern European–style social democracy and North American corporation-friendly neoliberalism. But landfilling was not the destiny of North Americans. Some Canadian provinces and U.S. states recycle more than Michiganders do, and some have invested in energy-from-waste infrastructure. If Michigan were to do that, it might still import waste (as have Sweden and Germany to fuel their waste-to-energy facilities), but at least this would seem to contribute to the public good, satisfying local demands for heat and electricity, rather than merely lining the pockets of waste firms.

But there are good reasons to be skeptical of Sweden's claims, given that the figures cited include only household waste disposal and not commercial, agricultural, or nuclear waste. When it comes to the latter, Sweden is understandably committed to a policy of permanent containment rather than reuse, as are nations throughout the world.[2] Even when it comes to household waste, however, processes officially designated as recycling may actually be a different form of disposal. Sweden's low rates of landfill are largely due to the use of thermal waste treatment, popularly known as "incineration." Incinerators are still controversial technologies and have to be carefully regulated and designed to lessen air pollution. This becomes more difficult when *anything* could be in mass waste. Partly as a result of irregular feedstock, the incinerator ash that remains may need to be landfilled, as was done at Four Corners. Some zero-waste advocates decry the co-optation of their slogan by incinerators and landfills, pointing out that energy-from-waste technology does not discourage wasting per se—rather, it leads to a reclassification of waste as a new source of private profit and public good. Moreover, energy-from-waste technology cannot do away with the disproportionate allocation of risks and opportunities associated with mass waste separation. As in North America, some Swedes live closer to landfills, nuclear storage facilities, incinerators, or recycling centers and some farther away; some work at them and many do not.

Addressing the problem of mass waste is about more than making improvements to, or recognizing the unintended consequences of, waste infrastructure. More fundamental change would require altering how people expect to live and not only how their waste is sorted and treated by others. But why should ordinary people have to change at all? Many environmental proposals and initiatives focus primarily on the actions of individual consumers and households. In the case of waste governance, this is clear in pro-recycling and anti-littering campaigns, both of which shift responsibility away from the sphere of production. Yet the vast majority of mass waste produced in North America (and therefore the world) is not made directly by households, but rather by industries that produce, package, transport, and sell raw materials and goods. If this is so, perhaps waste reform ought to target corporate, rather than consumer, practice.

Consumer-oriented waste reforms are clearly insufficient and possibly unjust. At the same time, it is important to recognize that those who produce and those who consume are as interdependent as those who manage waste and those who make it. Mass waste is, by definition, a collective matter. To oppose corporate profit and consumer preference is to imagine discrete spheres of social life (for instance, a home and market that are absolutely distinct), rather than the open-ended totality of a *waste regime.*[3] It would be wrong to blame either corporations or consumers alone for obsolescent electronics and bundling waste—neither would be filling the world's landfills were it not for the collective project of mass consumption.

Whatever we make of the Swedish counterexample, one could still propose other unrealized alternatives to North American or Northern European dump regimes. These alternatives would ideally challenge mass waste separation itself. One could imagine every subdivision, neighborhood, or town caring for its own waste and doing away with the problems of unevenly distributed risks and benefits. Landfills might be redesigned as interim locations for waste for which a use has not yet been found.[4] Such end-of-pipe solutions might increase the value placed on scavenging and tinkering, which would become more worthwhile practices to learn and teach, thus reducing social and personal investment in shopping for new goods.[5]

Any of these changes could alter what waste we produce or dispose of—not just what is done with it when we do—and none is beyond the realm of possibility. But I want to insist that with or without landfills, there will never be zero waste.[6] In principle, there is nothing wrong with using ambitious, utopian goals or simplistic slogans to get people to take action. But the appeal of "zero waste" comes from its peculiar combination of absolute calculability (zero as a definite number among other numbers) and an imagined escape from the realm of calculation as such (zero as a point that, once reached, marks an end to counting). In some ways, this has been part of the dual meaning of "zero" in mathematical discursive practice for centuries.[7] Just as old as zero, and deriving from similar cultural and historical antagonisms, are critiques of calculative or economic activity as inherently base, selfish, and morally suspect when compared with more transcendent values like love, self-sacrifice, and spirituality.[8] People derive enjoyment from the pursuit of zero waste partly because they imagine our deliverance from instrumental rationality and practical accounting as such.

From this perspective, the virtual fantasy of zero waste is not opposed to the base reality of mass waste—the two are mutually constitutive. It makes sense that "zero waste" would appeal to people who rely on routine separation from their own waste, who see waste regularly vanish from their toilets, their garbage cans, and their lives at no immediate cost, and prefer it that way. To maintain our treasured forms, something else needs to be disposable; the farther away this happens, the better it sustains the fantasy that our most precious values cost nothing, that they exact no toll on the world around us. Against this fantasy, we can and must affirm that there will always be some irredeemable remainder that cannot be reused or repurposed, however small a fraction of the whole this represents. Socially responsible consumers and corporations, carefully trained domestic waste producers and sorters, and closely regulated waste sites and facilities will not change this reality.

The impossibility of zero waste does not necessitate the use of landfills, but it is what landfills ultimately represent—a material limit that challenges all of our clever designs and environmental goodwill. Landfills are bad to think about, in part, because they are a testament to our inability to finally bend all of material reality to our collective will and, in so doing, finally escape our debt to the world.

The alternative to zero waste is not nihilistic despair at our lack of control, but humble acknowledgment that *both* the desire to endlessly accumulate and preserve new stuff *and* the desire to save the Earth from runaway accumulation are beset by real, material limits. The best-known image to symbolize waste is probably the icon for recycling: three green arrows that depict the circulation of matter as a closed circle that leaves no remainder. In modern waste governance, similarly, it has become commonplace to represent waste regimes by means of a pyramid or waste hierarchy. Closer to the peak of this hierarchy might be more favored strategies like waste reduction, recovery, and recycling; at its base would be the least-favored practice of disposal, upon which this book has focused. As is common, Sweden's Avfall Sverige represents landfilled waste—standing in for the abstract constraint of an irredeemable remainder—as an ever shrinking fraction of the waste regime. As the proportional use of alternative waste-management strategies increases, the amount of landfilled waste lessens, falling toward the hoped-for vanishing point of zero.

Instead of a hierarchically ordered pyramid or arrows moving in a perfect, harmonious circle, with their false promise of technically delivered transcendence, we need images of our relationship to waste that acknowledge the limits of control. Rather than imagine that irredeemable remainder of disposable excess as a leftover to be eradicated in pursuit of perfection, what if we considered it instead as the truth of the entire system, that which accounts for both profligate wasting and the appeal of zero-waste fantasies? As I have tried to show, waste is not a passive leftover but an absential force in our lives, not an object of technological mastery but something that represents our very failure to master things and surroundings. Orderly pyramids and closed cycles do not convey this idea.

Here is my substitute image: irredeemable, disposable waste is more like the core of a planet. Unseen and volatile, the center of the Earth is formless and fluid, dangerous and unapproachable. Yet it also provides the conditions for our planet's continued existence, its reason for lasting. The core is responsible for the magnetic field that blocks solar winds—much as the elimination and collection of formless mass waste ensure that things, people, places will last. Hidden away, a necessary fraction that maintains the outer layers of the world that we perceive and value, the disposed and despised core of excess waste keeps everything spinning and can never be eliminated. But it's easy to forget it's there.

There is no way to entirely opt out of some form of distributive wasting. The point is not that trying to change is pointless, but that the ascetic guilt of some environmental proposals and slogans is really the counterpart to arrogant and anthropocentric consumption, insofar as both individuate the relationship between waste producers and the world that surrounds them. It is as if a healthy, sustainable planet can be achieved only if we take it upon ourselves, as individuals, to leave no evidence behind of our presence on the Earth. The actions of the ideal environmentalist would thus resemble the self-denial of an ascetic monk, one who opts out of the balance sheet of economic activity entirely for the pursuit of higher ends. It's as if, instead of benefiting from what is absent, we should suffer from it—go without to lessen the trace we leave behind. If the medieval scholastic monk embraced absence to get closer to the full presence of God, then the ecologically aware, guilt-ridden environmentalist does so to discover a fuller, more authentic sense of self, one that seeks to overcome their antagonistic relationship with an impersonal and vulnerable Nature. But if zero waste is impossible, then leaving behind a trace is also inevitable. This irredeemable trace is, in fact, a core of who we are and what we value, even if we think we value nothing. There is no opting out of environmental impact: even Saint Jerome and Al Gore have to defecate now and then.

How might we transform, not just where mass waste goes and what we do with it, but how we represent and relate to it? Medieval monks used skulls and hourglasses, known as *memento mori,* as material reminders of the inevitability of death and decay, in order to live more pious lives (Rotman 1987: 68). Today, the subtraction of waste and its disposal elsewhere powerfully shape how we relate to transcendence and transience. Today, our everyday interactions with transience are structured by the continual filling and emptying of toilets, garbage cans, Dumpsters, and recycling containers. We know what they do, and we've seen in this book what becomes of their contents, but what subtle messages do such vessels communicate to us? How do they potentially distort our senses of impermanence and permanence, of the past and future, of our relationship to the material world? In a way, they are the opposite of skulls and hourglasses: their routine emptiness, their reconstructed sameness, communicate to us that they can be filled with what we reject, again and again, with no consequence. Instead, we need to find ways to resocialize and repersonalize our reliance on waste infrastructure.

One problem with any simplified sign standing for our relationship to waste—pyramid, cycle, or planetary core—is that it can potentially obscure the active roles of those who are implicated in mass waste separation and who shape it in turn. Acknowledging irredeemable waste as a core of who we are does not mean that we can disregard the decisions and actions of waste firms and their employees, regulators, and activist opponents. To the contrary, if waste is an inevitable and necessary force in our lives, we ought to care about what becomes of it rather than leave it to other people and places to care for it on our behalf.

I have argued that waste management is a social relationship, rather than an unmediated connection that individual consumers have with an abstract and impersonal Nature. My former coworkers would occasionally acknowledge the care they provided for others, but their day-to-day struggles were largely individualized, their collective contribution dissimulated by highly personal experiences of disgust and denigration. The possibility of recognizing mass waste separation as a social process, and hence something for which we are collectively responsible, was foreclosed from landfilling by design. What if we were to have landfill workers care for our remains more directly? What if their labor were more visible and its object unforgettable?

Here is my policy recommendation: let us bury our dead in landfills and not graveyards. In fact, let us merge the two. In death, let us reconnect with the waste we spend our lives refusing. For some, this could be relatively easy to do. Their cremated ashes need only be discreetly added by their next-of-kin to the weekly garbage pickup—maybe this has even been done before, countless times. But one could also imagine the burial of human beings being done with the same ceremony currently reserved for ATF-monitored burials of smoke-damaged liquor. What if Bob had to stop dumping everyday for a funeral, at regular intervals? This would mean the repetition of that revelatory scene, discussed at the start of this book, caused by a Canadian driver's unfortunate death. All attendees would bear witness not only to the ritual, but also to the many trucks lining up with their loads, waiting to dump, the many machines and workers required to make our collective remains vanish. All waste workers, similarly, would be directly confronted with the people whom they care for in secret, people whose absent presence is otherwise encountered only through the medium of their discards. Bob might gain more public recognition as a manager/mortician; his actions would also come under more scrutiny, as happens to those who manage North American burial services generally.[9] In fact, this might help Bob and his fellow managers cultivate greater sensitivity toward others, a quality their employees occasionally found lacking.

If being buried in a landfill seems absurd or even offensive to contemplate, it is helpful to consider why. Reviewing the possible objections one might expect is an exercise well worth engaging in, because it reveals the barriers that currently prevent North Americans from relating to our mass waste and that foreclose the possibility of its social reclamation. One might object that if this practice were generally adopted, it would reduce the productivity of landfills. Some of my former coworkers and employers might hold this view. But it is not uncommon for special forms of disposal to delay or focus their activities. When the ATF arrives, they have no choice but to surrender such concerns for the collective good. It just so happens that it's far easier to marshal state support to ensure the proper disposal of smoke-damaged alcohol. That we might lose faith in the ability of new commodities to satisfy our desires is seen as a significant risk worth addressing; while developing opportunities for us to temporarily dwell upon (or eternally dwell within), the material consequences of this phantasmic drama of consumption are not considered worthwhile.

A more practical objection might concern whether landfill burial, like cemetery burial, can actually serve as a sufficient medium of social expression. Cemeteries are similar to landfills in some respects, insofar as they are designed to downplay their materiality—all evidence of the decaying corpse, for example—so that loss and transience are mediated through signs of permanence. But one glance at an average North American cemetery reveals the social differentiation of contemporary headstones. Although sites like Arlington National Cemetery are no less socially meaningful for being highly uniform, this takes away the ability to encode particular messages about the deceased beyond very general attributes. Landfill design would certainly have to be reimagined to do something it was never intended to accomplish—bury while also preserving a memory of loss. With new GPS technology installed in the cabs of compactors and bulldozers, the precise location of particular waste loads can be virtually mapped onto any landfill with far greater precision. Bob claimed that they would soon have the means to locate mistakenly discarded, precious objects should the need arise again, as it has in the past. Should one wish, it would be possible to know where remains lay and even mark the spot on the surface. One could even buy a burial plot in the landfill's future air space as is currently done in cemeteries, or decorate the surface once it has been capped, to signify the particular location of a loved one.

A much stronger moral objection would be that landfilling human remains would insult the memory of the dead or threaten religious ideals. To combine landfills and cemeteries, according to this view, would be to irreverently mix what should be kept separate. In fact, such an impossible mixture has already happened and the result was predictably traumatic for those affected. After the terrorist attack in New York City on September 11, 2001, recovery efforts began to find those who had been buried in the rubble of the twin towers. The retrieval of bodies proved difficult and many were never recovered, which made it that much harder to properly mourn those tragically lost. When the rubble was then relocated to nearby Fresh Kills landfill for further sorting and disposal, it was as if those missing were being landfilled like so much meaningless trash.[10] In addition to their polluted and polluting insides, landfills represent a medium designed for forgetting. All that is entombed within becomes nothing more than undifferentiated material. Even as the particular biographies of its insides are forgotten, furthermore, a landfill lasts as a permanent fixture of the landscape. It is, in this sense, the complete opposite of a proper graveyard, which is ideally differentiated into singular plots of pure memory.

Landfilling a loved one would seem to deny the singular importance of those we lose to death, the special status of their corporeal remains as something other than a mere thing. Put simply, a human corpse is sacred, and our mass waste is the very definition of profane—that is, unimportant, insignificant, merely material and not at all spiritual or holy. But what does it mean to perceive waste as mere things in contrast to the not-quite-thing of a human corpse? At face value, it would seem to represent a way of preserving human dignity and value by maintaining our distinction from the rest of the living and nonliving world.

But there are many people, throughout the world and human history, who have buried their dead with their discards. They do so for the dead as well as the living. Materials may be wasted through burial itself. Consider "grave goods," which are valued items left with the deceased. This practice is used to signify the status of the dead, as well as to avoid property disputes among the living. In some cases, the placement of the dead may serve more practical functions for the living. Just as landfills initially spread in North America as a way of providing material and structural foundation for new buildings, so have the dead been used for similar purposes. There is archaeological evidence of the bodies of children and infants buried underneath bridges and buildings throughout the world, presumably to serve as a source of spiritual foundation and protection. Corpses may even become a scarce resource in contexts where death is necessary for the productive regeneration of life.[11]

The deposition of bodies and materials may be closely entangled in other ways. The Endo, a subgroup of the Marakwet people of western Kenya, have been known to choose the location of human burials to signify connections to waste items and the material practices they represent. Here, burial of the dead and waste management are mutually constituting: women ought to be buried with the chaff and old men with goat dung. This is a way not of devaluing life, but of protecting it—to bury people otherwise, for the Endo, would be to endanger the living.[12]

A similarly intimate relationship between objectual and human remains is evident among the historical and contemporary peoples of the North American Southwest.[13] Pueblo burial in the twentieth century is thought to be relatively continuous with prehistoric burial practices in the Southwest. In both cases, waste disposal does not represent a separate sphere of material practice distinct from sacred and ceremonial activity. At the same time that Jean Vincenz was experimenting with his new sanitary-landfill design in Fresno, California, in the early twentieth century, less than a thousand miles away the Pueblo were doing a very different form of disposal. While Vincenz was seeking to reduce the number of people present at a dumping site to increase efficiency (by eliminating scavengers, in particular), Pueblo were peopling their dumping ground even further. One might even suggest that waste does not exist in Pueblo cosmology, only corpses—the husks of once animated forms, endowed with spirit. People and things are not equal in stature, necessarily, but they are united by their common fate: entropic decay and regeneration await all materials and beings. For this reason, dust or ash can serve as a representation of the continuity underlying dissimilar things—a medium of the spiritual power of Our Ancestors.[14]

According to tales that have been told to Pueblo children, piles of discarded materials are inhabited by Ash Boy, described as a manifestation of the Supreme Being and as part of the mythological complex of the popular cultural hero Poseyemu. Connected to the hearth and the home, Ash Boy exists in the ashes from fireplaces, which are regularly cast off at village dumps along with the other dust and debris swept out of living spaces. Ash Boy has presided over cyclical ceremonies that rededicate sacred kiva rooms and played a role in local mythology. Through Ash Boy, Pueblo form moral relationships with their waste. This does not mean they stop dumping it, but it radically changes how they work and live with what they dispose of. At the same time that Vincenz's landfill design was spreading throughout North America, Pueblo children were having ashes from the hearth placed on their foreheads to summon Ash Boy's protection.

The influence of indigenous American beliefs on North American environmentalism is well known—most famously (or infamously) in the "Crying Indian" commercial that first aired on Earth Day in 1971, financed by corporations seeking to represent waste as primarily the responsibility of individual consumers.[15] But following the Pueblo example would not mean reenchanting the Earth in the abstract and our individual relationships to it. By collapsing the separation between waste producers and waste deposits, Pueblo ash mounds become reintegrated into everyday life and cosmological reckoning. To learn from them, we need to recognize and affirm waste disposal as a social relationship, to express and maintain moral and social bonds through our discarded waste.

Waste infrastructure in North America has been purified from any association with the death and disposal of human beings, from spirituality or the sacred. Landfilling is represented as a profane exchange of material, with economic and physical—but no moral—significance. Neither Bob nor Vincenz considered themselves ritual practitioners, but neither have the Pueblo, necessarily.[16] Both the sanitary landfill and Pueblo ash mounds developed as strategies to manage collective discard for the sake of absent beings—whether waste producers or spiritual ancestors—located in some virtual elsewhere. And there is no reason to assume that Ash Boy is any less technically innovative than a Vincenz landfill. The Pueblo had experienced centuries of violence, domination, and outmigration prior to ethnographic and archaeological investigations into their practices. Is it surprising they were so invested in the material remains left behind by former kin who had died or left the settlement behind? Vincenz's technical innovation was really a combination of familiar techniques—the practice of dumping going back millennia as well—and his contribution was as much social as it was technical. Both the sanitary landfill and the ash mound involve immanent communion with materials and transcendent separation from them. The most important difference, for the purposes of my argument here, is that landfills do not promote new social relations between waste producers and waste sites, but foreclose them entirely.

The waste practices of the Endo and Pueblo force us to consider that objections to burial-at-landfill are historical and mutable. If we refuse such an option, it is in part because we have acquired a compulsion to remain disconnected from what we discard and all that it represents. The compulsory refusal of impermanence that landfilling has encouraged also desocializes our experience of this process, foreclosing recognition of the care others provide on our behalf. Landfills are a place where things go to be forgotten and, having been forgotten, they amass into something more monstrous and impersonal than any graveyard. In some ways, landfills have become so alien and removed from our lives that they've acquired a quasi-sacral character. I have discussed how the fear and alarm they inspire in others can far exceed the material threat they might actually pose, leading landfill opponents to speculate about the intentions of foreign truck drivers as if they were the embodiment of the waste they haul (see chapter 5). Even the regulatory principle, enshrined in RCRA, that anything that comes into contact with garbage therefore becomes garbage—as when leachate spills on the ground—is an impossible rule to implement, though it may make sense as a precautionary measure against contamination (see chapter 1). In this respect, landfills are not unlike coffins for our mass waste, not only disposing of our discards but also entombing them, keeping the undead from rising again. And this makes it harder to reclaim or reform mass-waste infrastructure.

Vincenz's vision of disposal shaped North American life throughout the twentieth century, and that cannot be undone. Even Pueblo are now routinely exposed to contamination from the nuclear-waste landfill that surrounds the Los Alamos Nuclear Laboratory (Masco 2006). Whether our preferred method of final disposal (for all that irredeemable waste we cannot recycle or reuse) will be landfill, incineration, or something altogether different remains to be seen. Whatever technical modifications are made to mass waste management, it is important that we find ways to get closer to our sites of final disposal, to resist and reimagine their separation from our lives. Combining landfills with cemeteries is my (impossible) solution to the (unrecognized) problem of our collective separation from mass waste, of the distorting image of social and material reality it leaves us with. We need to change the meaning of places like landfills and the labor done there, to make mass waste not only newly visible but newly meaningful. Better said, we need to recognize the meaning it already has for us. And if we are so bothered by having mountains of trash as the lasting memorial to the lives we lead, we should lead them differently. But only by wrestling uncomfortably with who we really are, with what seems (im)possible and why, can we actually change.

LANDFILL FUTURES

Landfills are part of the future of North America. Even if they all close and new waste relations begin to emerge, they have irreversibly transformed our lives and landscapes. In a sense, we are stuck with them. Some writers have drawn on this troubling relationship as a resource to imagine possible and improbable futures.

In a short story called "Landfill Meditation" in a book of the same name, Ojibwa poet and trickster storyteller Gerald Vizenor reclaims landfills in this way. He introduces the reader to Martin Bear Charme, a man who made his fortune dumping waste into a worthless wetland, which he has now turned into a site of spiritual meditation meant to reconnect people to the earth. Landfill meditation restores this lost intimacy with the earth, Charme says, because it reconnects people with their abandoned waste: "We are the garbage, the waste, we make it and dump it, to be separated from it is a cancer-causing delusion" (Vizenor 1991: 104). Here Vizenor playfully subverts the racial economy in settler societies whereby indigenous and minority populations receive waste that others only too readily give them.

In *Infinite Jest* (1996), David Foster Wallace offers a complementary reflection on waste and border politics in North America. In a strange future, much of the northeastern U.S.–Canadian border, including the (largely rural) northern half of New York State and some of Ontario, has been transformed into a massive toxic dump, known alternately as the "Great Concavity" or "Great Convexity," depending on the side of the border one is on. This vision of environmental destruction imagines the permanent transformation of a rural elsewhere into an appalling site of anti-nature. The environmental impact of this abomination is largely absorbed by Quebec and, in an absurd reversal of colonial history, Wallace has the United States cede this wasteland to Canada. Wallace's fantasy mocks both the territorial border politics of North America and the market in waste services that unites them.

In different ways, Vizenor and Wallace play with the complicated relationship between waste producers and their waste. In these homologous visions of a wasted future, what North Americans dump offers opportunities to reaffirm a part of ourselves we have cast aside and/or cede responsibility for it entirely. As I have argued, continuing to choose the latter option affects not only the planet in the abstract, but actual people who continue to care for and about our waste so that we don't have to.

One day, Four Corners landfill will close. At some point thereafter, the site will be officially reclaimed by Harrison Township. Reclamation is an important part of landfill–host community agreements. Money is set aside by landfill companies to prepare for the process, which may involve improvements to the site to make it safe and reusable for other activities. During my research, North America's leading waste-disposal company sponsored national television commercials and newspaper ads to convey how closed landfills can benefit communities, by being turned into golf courses, baseball fields, or wetlands. One landfill in southeastern Michigan is being shaped in the hope that it will become a new ski resort once capped. This idea is not new. "Landfill construction"—prominent waste firms have used the idea of reclamation to challenge the popular conception of the landfill as worthless and harmful to people and the environment. But things are rarely so simple in practice.

Reclamation has already begun with Harrison's first landfill, CSG. Despite various environmental leakages over the years, members of the Harrison High Flyers Club have met regularly on top of the landfill—the second highest point in town after Four Corners—to fly remote-control airplanes together. Detroit's swat teams have also practiced exercises at CSG, presumably because of its seclusion and heterogeneous terrain. At one corner of CSG, a thick mound of dirt serves as a shooting range, which the local police department occasionally makes use of. And every fall, Harrison holds its annual "Turkey Shoot" there, where local families shoot at (inanimate) targets to win small prizes. But CSG and Four Corners are very different sites. The former looks like a series of small hills. By the time Four Corners is complete, several decades from now, depending on how quickly air space is added and filled up, it will be more than five hundred feet high and will span one square mile. After it is capped, or covered with sufficient soil, the only regular activity left will be for laborers to mow its slopes and for gas technicians to monitor the power plants, which will operate for years afterward as the waste slowly putrefies within. A small, gassy, grassy mountain, Four Corners may not provide the same advantages to the community that its predecessor now does—it may even be closed to visitors.

And yet the people working and living in the vicinity of Four Corners are already finding ways to make the site their own. In the first three chapters,

FIGURE 8. Todd at a retention pond.

I showed how landfill employees struggle with disgust and indignity in various ways but also attain values derived from the landfill. Those who live near the site also make it meaningful, beyond engaging in the political demonstrations discussed in chapter 5. A former engineer from the site now lives across the street with his family and dogs, strays they routinely find wandering the area. He, his wife, and child used to wander onto the landfill territory in the winter, explore the forest and wetlands, and skate on the frozen ponds. This stopped once the fence was put up, however. The fence was erected as a response to a fire up top that destroyed two tippers, which many of my coworkers believed was started by a beleaguered former employee. The fence is, in a sense, a barrier to unwanted and illegal forms of local reclamation. Even with it there, road signs around the site were repeatedly vandalized. And other neighbors still requested cattails from the landfill's perimeter ditches, which they were allowed to take and use for decoration and display.

Once landfill operations cease, it is certain that more forms of reclamation will emerge as neighboring residents continue to make the site their own through various means. But ultimately, Four Corners was designed more for containment than for reclamation. Like other landfills it is, above all, a place for the storage of polluted and polluting materials. In fact, there is a paradoxical sense in which pits of transient matter are themselves *permanent*. Just as Bob witnessed firsthand how waste circulation is bigger than one man's life, Four Corners landfill will likely remain a store of negative value long after the people, places, and possessions it helped maintain lose what formal durability they once had.

If things only ever last temporarily and transience is the rule, then these material conditions, while equally real, are perpetually out of balance or asymmetrical. By facilitating disposability, landfills go against the current of entropy, reversing the imbalance and allowing us to cling to familiar conditions in what are otherwise challenging and uncertain times. The glimpses of permanence we get are no illusion. As Zygmunt Bauman writes, a sense of eternity is experienced every day in our encounters with "faces and places, routines and rituals, sights and sounds which are familiar, stay familiar and are expected to remain familiar as they are now" (2004: 104).[17] The act of tossing something into a garbage can, recycling bin, or Dumpster scales the ravages of time down to size: what we lose is truly insignificant compared to what we keep and maintain. But the act of landfilling amasses waste at mountainous scales out of sight, and what remains is far bigger and longer lasting than even the people who built it or supplied it with material. Like those they serve, landfills are built to last. As if in an unending spiral, what was once permanent becomes transient and what was transient becomes permanent.

I have not been back to Harrison since leaving Michigan in 2008, and cannot say for sure what has become of all the people I knew there. I do know that Four Corners is still there, bigger and more imposing than the last time I smelled it or walked across its slopes picking paper. Without knowing who still works there or why, as I write this I know that, for now, the garbage keeps coming.

The garbage keeps coming.

The garbage keeps coming.

NOTES

INTRODUCTION

1. All names of local places, persons, and businesses associated with this landfill have been changed. This was a condition of my being allowed to conduct research at Four Corners as an employee. Where necessary, I have done my best to omit or alter inconsequential details in order to honor this obligation.

2. For a discussion of "being affected" as a way of participating in a research setting, see Jeanne Favret-Saada's *Anti-Witch* (2015), which argues that an ethnographer's intense affective experiences should be accorded epistemological status, difficult though they may be to represent.

3. Using the term "North American" to depict only the people of the continent's two most powerful countries is admittedly problematic, much like the use of the hemispheric term "America" to depict only the United States. While I use this common toponym throughout this book for ease of reading, my main focus is on a landfill composed of waste from Canada and the United States, and not from Mexico. Most of the people who appear in the text may be from the United States, but most of the waste discussed is from Canada, so the latter are indirectly but significantly part of the action.

4. In the United States, Type I landfills are for toxic waste and Type III landfills are designated for construction and demolition waste.

5. Various authors have looked beyond individual waste sites or particular waste streams to grapple with the much bigger problem of American waste in general, including Alexander (1993), Rogers (2005), Royte (2006), Kennedy (2008), Thomson (2009), Clapp (2010), Humes (2012), and MacBride (2012).

6. On the concept of waste regime as a way of exploring the representation, production, and management of waste as a totality, see Gille (2007).

7. On histories of cleanliness in North America, see Hoy (1996), Tomes (1999), and Brown (2009).

8. Douglas, Leach, and Dumont credit different mechanisms for creating and resolving the problem of ambiguous and abject phenomena (see also Kristeva 1982).

But whether one prefers a cognitive, linguistic, ideological, or psychoanalytic explanation, each privileges uniquely human schemas for making the world meaningful, while downplaying the role of the material world in shaping representational practice. See Reno (2015) for a longer discussion of contemporary discard studies as an alternative to these structural and symbolic approaches. For a discussion of the all-too-human and the limits of anthropocentric theories of representation, see Kohn (2013).

9. Deacon (2012: 194) suggests abandoning "descriptive notions of form" associated with prevailing conceptions of order and disorder; in their place, he argues that we should think about "regularity and organization in terms of possible features being excluded—real possibilities not actualized." Douglas (1984: 95) makes a related observation: "Order implies restriction; from all possible materials, a limited selection has been made and from all possible relations a limited set has been used." However, it is important to recognize, contra Douglas, that all the order found in the world is not derived exclusively from humans actively constraining or allowing for possibilities.

10. In different ways, Derrida (1976) and Deleuze (1994) both demonstrate how the repetition and presence of self-identity are constituted by difference and alterity (see Patton and Protevi 2003). Similarly, Reno (2009), Bennett (2009), and Gregson and Crang (2010) use waste to highlight material becoming and the inherent instability of all forms. But analytically privileging processes of deformation, in discard studies, risks downplaying the constraints on deformation that are introduced through the subtraction of waste.

11. The idea of form I use here is therefore different from the idea of form as an intelligible object, opposed to formlessness or indefiniteness (see Staten 1984: 5–7). When I mean "formless" in this latter sense, I generally use the term *indeterminate*. Douglas (1984) and Leach (1964) tend to conflate these two senses of form, but they rely far more on the latter sense to discuss pollution and taboo. A phenomenological defense of forms as actual, dynamic objects, rather than mere ideational projections, can be found in Harman (2005, 2009).

12. Severin Fowles (2010, 2013) argues that the importance of absence is often neglected in discussions of materiality and thing theory. We normally think of causality as involving things in immediate contact with each other, as in Newtonian physics. But what is missing or absent can have causal consequences by limiting what is actually available or copresent in a situation (see Juarrero 1998, 1999; Deacon 2012; Connolly 2013; Kohn 2013).

13. Hurricane Katrina had recently been in the news, and Eddy expressed disgust that people were homeless in the Gulf region while we were destroying homes in Michigan. For more discussion of this incident, see Reno (2012: 269–70). Shannon Lee Dawdy (2010: 774) notes a similar process during the disaster-recovery effort after Hurricane Katrina.

14. "Reconstructed sameness" is Joseph Masco's (2006: 7) gloss for Walter Benjamin's critique of the industrial forms that permeate modern life. Benjamin aimed to undo the beguiling effect of our everyday encounters with mass-produced

commodities destined for obsolescence (see also Dawdy 2010). In a complementary analysis, Greg Kennedy (2008) argues that the routine elimination of waste has deprived the things that surround us of being, making them seem almost ghostlike. Arguably, landfills encourage a reduction of being to lasting—that is, they lead us to equate existence with staying the same, with transcending the world rather than dwelling and surviving within it.

15. My use of biological and cybernetic analogies to explain waste disposal is not intended to naturalize or dehistoricize industrial capitalism, but precisely to highlight the noncapitalist and even nonhuman processes upon which capitalist political economy subsists (see also Kohn 2013: 153–90). I follow Gibson-Graham (1996) and Graeber (2011a), who argue that economies always contain diversity, which prevents them from ever being uniformly capitalist or noncapitalist. On one hand, as Crooks (1983, 1993) shows, waste management has been thoroughly privatized and overrun by corporate interests in Canada and the United States. On the other hand, capitalism in North America is both made possible and threatened by mass waste disposal, aspects of which predate industrial capitalism entirely. On the history of the relationship between capitalism and waste, both as a provocation to privatize what is held in common and as a limit to profit, see Foster (2002), O'Brien (2007), Gidwani and Reddy (2011), and Gidwani (2013).

16. According to Latour (1993), becoming modern is not a totalizing historical transition but an impossible fantasy, given that it hinges on a purification of reality into incommensurable domains, like Nature and Society. This paradoxically results in a multiplication of hybrids that illicitly recombine the two. Landfills are just such a hybrid, which is one of the things that makes those who rely on them moderns.

17. Elias (1994), Bakhtin (1965), and Stallybrass and White (1986) analyze the relationship between social and material disorder as a source of social differentiation throughout European history. Weiner (1992) and Graeber (2007, 2011b, 2013) provide accounts of various efforts throughout the world to achieve distinction through material transcendence. One successful means of transcending process and change, for example, is to be immortalized as a quasi-mythical or godlike being altogether beyond the mundane world (Sahlins 2004, Kohn 2013: 180–1). One can also achieve transcendence in the opposite way: by embracing transience and bodily continuity with the world (Parry 1982: 94).

18. Throughout the Global South, a lack of funds for waste and wastewater management infrastructure is often related to structural adjustment arrangements with powerful lenders like the International Monetary Fund and World Bank, as well as broader imperial formations that have deferred the promises of modernity through perpetual ruination (see Chakrabarty 1992, Stoler 2013). Bauman (2004) and Yates (2011) argue that modernity and global capitalism result in the proliferation of human waste, as human lives are devalued and made more precarious. Indeed, in this sense, the ruination of uneven development is not exclusively distributed according to divisions of Global North/South or Core/Periphery. As I discuss in chapter 4, it has also shaped post-agrarian and post-industrial landscapes of the midwestern United States.

19. The concept of biosociality assumes that relationships exist between humans and nonhumans that are dynamic and corporeal (see Ingold and Palsson 2013). On waste processes as multispecies affairs, see Hird (2012) and Reno (2014, 2015).

20. Modern representations of entropy are often framed in terms of risk. Some scholars have regarded the late twentieth century as a period of transition, for modernity, toward a global risk society wherein the calculation and management of uncertainty begins to dominate everyday life as well as national and supranational power in the way that political economy once did (see Beck 1992, Adam 1998). Given that waste management has played a constitutive role in urban governance and economy for millennia, however, it arguably represents an older type of risk sociality and governance, one that calls into question divisions of world history into fixed and stable epochs.

21. Masculinized waste work contrasts sharply with the more often feminized and lower-paid waste work of cleaning, child and elder care, and informal recycling. On informal recycling as a gendered and marginalized practice throughout the world, see Medina (2000). On the similarly gendered labor of elder care, see Jervis (2001) and Twigg et al. (2011).

22. Bill Maurer describes homologies as "samenesses whose difference the act of substitution restages" (2005: 13). One could characterize my coworkers as "para-ethnographers," after Holmes and Marcus (2005), in the sense that their movement in and out of a landfill's orbit parallels the dislocation associated with ethnographic practice and evokes homologous reflections of similarity and difference, home and away.

INTERLUDE: A NOTE ON DRAWING AND ETHNOGRAPHY

1. This also allowed me to creatively secure anonymity for my informants, as I had promised them, whereas photographs would not.

CHAPTER 1: LEAKY BODIES

1. Alexander and Reno (2014) explain attempts to replace landfills and fossil fuels in Great Britain as a form of biopolitics.

2. The "sanitary idea" popularized by Edwin Chadwick in England was a relatively simple policy designed to improve public health by separating people and dwellings from transient, miasma-producing materials. According to David Armstrong, the new sanitary regimes that emerged in the nineteenth century "monitored a line of separation between the space of the body and that of its environment" or "anatomical from non-anatomical space" (2002: 7–8). If sanitation is premised on the division of the corporeal from the environmental, it also assumes bodies that can be thought of in isolation from their surroundings and supports the professional

division of medical doctors from sanitary engineers. On the transition from a sanitary to a medical model with the rise of bacteriology in the United States, see Porter (1999: 159–61).

3. As Michel Serres writes, "the organism is a barrier of braided links that leaks like a wicker basket but can still function as a dam" (1982: 75).

4. A recent appraisal of the hygiene hypothesis, in the context of rural Tanzania, can be found in Wander et al. (2012).

5. In her in-depth study of Danish hospitals and clinics, Annemarie Mol (2002) demonstrates that any patient's body constantly changes, has so many components, and exists on so many scales that it is never absolute or fixed, but only ever hangs together across distinct sites of medical practice. The body's multiplicity is not an impediment to medical practice. Rather, it enables pathologists, surgeons, and therapists to approach phenomena like arteriosclerosis using radically distinct, even incommensurable methods.

6. For discussions of the moral and economic significance of medical discards, see Collier (2011) and Halvorson (2012).

7. Transformations in waste practices are critical for changing ideals of selfhood, as is clear if one considers the historical shift from dark, unadorned outhouses to indoor toilets in locked, well-lit, and mirrored restrooms (Laporte 2002). Osborne (1996) and Joyce (2003) provide historical accounts that relate sewage infrastructure to liberal ideals of freedom and private ownership. In a similar way, Burrell (2012: 49) argues that dust-free, air-conditioned, and cosmopolitan Internet cafes provide the urban youth of Ghana with a sense of open-ended and indeterminate possibility, which contrasts sharply with the other places and spaces they occupy.

8. This is arguably a contemporary continuation of medieval Thomism: if the body matters only because it is inhabited by a person, the qualities of the person can be ushered forth, become manifest, only through the imperfect display of transient flesh. Following Foucault's critique of liberalism, Nikolas Rose (1999, 2007) has written extensively on the forms of self-government associated with modern biopower, including the growing influence of medicalization.

9. Using the body as a metaphor can also be problematic, as when weapons scientists describe nuclear bombs as living beings with natural life courses rather than as destructive devices (Masco 2006: 80). There is also a tendency to represent both artificial creations and social and political collectives as bodies, often in a gendered fashion that universalizes masculine bodies and marks women as exceptional (Martin 1987, Gatens 1996). The characteristics of a landfill body as I describe them (e.g., skin, internal organs, flesh . . .) are neither exclusively human nor, necessarily, sexed. Think of landfill bodies as asexual, like the stick insects and whiptail lizards that dwell in California, not far from where the first landfill was birthed in Fresno. A landfill body is not singular and unified, like the ideal Thomist body, but leaky and multiple. Moreover, one could just as easily imagine the body as a landfill: "I am not an agent but a hive of activity. If you were to lift the lid off, you would find something more like a compost heap than the kind of architectural structure that anatomists and psychologists like to imagine" (Ingold 2010: 17). Both a hive of unruly

activity and a controlled architectural design, each landfill body is an imperfect combination of local surroundings, waste, and human labor.

10. See Logan's (1995) account of the ecstatic skin of the Earth. For an alternative, early-modern view on the exchange of substance between decaying bodies and atmospheres, see Corbin (1986: 11–21).

11. According to a study by Klepeis et al. (2001), North Americans spend more than 90 percent of their time in enclosed buildings and vehicles, on average.

12. On the immanent sociality of engaging with one's surroundings on foot, see Ingold (2004).

13. Jablonksi (2006) provides a comprehensive overview of the biological and cultural significance of skin in humans and other animals.

14. Darwin (2005) and Kolnai (2004) are credited with some of the earliest and most influential interpretations of disgust. Miller (1997) and Rozyman and Sabini (2001) provide more recent overviews of disgust scholarship. One influential argument is that experiences of moral and physical contagion are reducible to the body's permeability to its surroundings, disgust being an adaptive method of disease avoidance (Rozin et al. 1986, Curtis et al. 2011). This would explain why physical revulsion is often concentrated around the bodily apertures: they are means of possible contamination from outside (Fessler and Haley 2006). This idea of disgust relies on the notion, often ascribed to Darwin, that the external environment is opposed to the individual organism's survival, something they must adapt to and overcome. For a critique of this view of organisms and evolution, see Ariew and Lewontin (2004). The very existence of apertures suggests a more nuanced relationship between inside and outside, and an altogether different understanding of survival, than many interpretations of disgust tend to suggest.

15. The idea of "dividuals" or "partible persons" is credited to Marriott and Inden (1977) and Strathern (1988), respectively. They juxtapose South Asian and Melanesian ideas of the person as consisting of relations to western, liberal conceptions rooted in individual self-creation and private property, as explained most famously by Macpherson (1962). Jonathan Parry (1994) argues that Indian Hindus (and, arguably, other dividualizing ethnosociologists) can and do see personhood in both ways.

16. For an explanation of affect programs, see Griffiths (1997) and Protevi (2009, 2011, 2013). According to Protevi, sub-subjective affects like rage or disgust are more amenable to manipulation and cultivation by supra-subjective forces and relationships, something that can be harnessed on behalf of an invading empire's war machine, for example (2013: 79).

17. Iacuone (2005) discusses the role of sexual humor in the maintenance of heteronormativity in the construction industry. Bob regularly used homosexual humor when conversing with his employees. When I observed this, it seemed to reassert his superiority in the class hierarchy, despite the fact that it was presented in the formal, metadiscursive character of friendly speech (i.e., as a way to level status differences).

18. The very word *house* is etymologically linked to both meanings of *hide*, the verb "to conceal" and the noun meaning animal *skins* (another cognate of *house*);

this is not surprising, given that animal skins/hides once served as construction material for homes.

19. Jackson (2011) makes a parallel set of observations concerning smell, relations to place, and environmental contamination in a First Nation and Canadian context.

20. Rathje and Murphy (2001) take issue with the notion that materials degrade within landfills. Some form of settlement and gas production are common in many landfills, however, and would not occur without some biological degradation within, no matter how slight.

21. Paraphrasing Steven Feld, who explores the importance of sound to experiences of place in Papua New Guinea, the importance of smell to the everyday life of my informants could be called an "[olfactory] epistemology of emplacement" (1996: 105). For a discussion of other ways in which bodies sense pollution, see Shapiro (2015).

22. On the importance of pleonastic pronouns in English for emplacement, see Bolinger (1973).

23. Rosen (1997) and Chiang (2008) describe the noisomeness of early-modern America. Armstrong (2002) contrasts the older policy of quarantine, which involves social division and uneven treatment, with the nineteenth century's sanitary regimes, which were meant to apply to the lives and conduct of the whole population. But waste management inevitably redivides, through the quarantining procedure of mass waste disposal, the public it unites through sanitary mass waste collection.

CHAPTER 2: SMELLS LIKE MONEY

1. "Smells like money" has also been observed among waste pickers in Sofia, Bulgaria (Elana Resnick, personal communication). I suspect that its use could be even more globally widespread, given the economy of the expression and its semiotic multivalency.

2. Waste workers must become what Susan Buck-Morss (1992) describes, following Benjamin, as anesthetic subjects: those who have become numb to the sensory stimuli around them. MacLeish (2012) draws on similar ideas to characterize the masculine unfeeling of American soldiers wounded or vulnerable in warfare.

3. David Pedersen (2008, 2013) offers an original and persuasive account of the relationship between value and storytelling. In this he follows Diane Elson's (1979) reinterpretation of Marx's "value theory of labor" as a critique of capitalist social relations rather than some kind of ahistorical economic formula.

4. In an influential ethnography written while working as a machine operator at a Chicago factory, Michael Burawoy (1979) explains how managers and owners try to coerce their employees to consent to their authority, including control over workplace narratives. One successful way of doing this, which I observed at Four Corners, was to pit workers against each other through the internal labor market.

5. Dirty work and stigma have been topics of sociological investigation for several decades, beginning with Hughes (1951, 1962) and Goffman (1963) and including, more recently, Ashforth and Kreiner (1999, 2014). The dirty work of waste workers is often assumed in this literature but not investigated closely (for notable exceptions, see Walsh 1975, Perry 1998).

6. On the new American middle classes and their relationship to spending and saving, see the classic study by Robert Bellah et al. (1986). Robin Blackburn associates this reflexive tendency—becoming more and more concerned with managing the financial dimensions of almost every domain of life—with a process of society-wide *financialization* (2006: 31). In this way, he emphasizes how fostering middle-class identity now depends more on relationships with "finance houses" and "commercial suppliers" than with the state or the community.

7. Harvey (2005) provides an overview of the history of neoliberalism as an ideology and social movement, and Connolly (2013: 58–60) of its failure to take into account systems of nonhuman self-organization beyond the market. Ong (2006) and Piot (2010) describe the complex reception and translation of neoliberal governance into Asian and African contexts, respectively.

8. As described by Blackburn, at the end of the first decade of this century, "Finance houses . . . teamed up with retailers to shower so-called gold and platinum cards on all and sundry with the hope of ratcheting up consumer debt" (2006: 44).

9. On the relationship between neoliberalism and the financial crisis, see Palley (2011).

10. Elizabeth Povinelli (2006: 4) argues that the conditions of late liberalism amplify an abiding tension between an autological, self-determining subject and a genealogical, person-producing society.

11. As Richard Sennett and Jonathan Cobb write, "[T]he class structure in America is organized so that *the tools of freedom become sources of indignity*" (1993: 30). Kathryn Dudley (1994: 128–9) argues that autoworkers in the Midwest may see education as a valid path, but not one that necessarily leads to happiness, intelligence, and success.

12. For a complementary and nuanced analysis of class ambivalence from the perspective of the upwardly and downwardly mobile children of working- and middle-class parents, see Bettie (2003).

13. According to Moishe Postone's analysis of capitalism as a form of social domination, "Abstract labour begins to quantify and shape concrete labour in its image; the abstract domination of value begins to be materialised in the labour process itself" (1993: 182).

14. This shift could be described as a form of neoliberal self-government (Rose 1999; Frederick and Lessin 2000; Rasmussen 2010, 2013) or as a "DIY Project of the Self," characteristic of reflexive modernization (Abbott and Kelly 2005).

15. E. P. Thompson (1967) produced the definitive historical account of the connection between clock time and conceptions of space and time within industrial capitalism. Time-discipline is part of the expropriation of worker knowledge and skill by the professional managerial class in general (Hetrick and Boje 1992: 55;

Abbott and Kelly 2005). Thompson's ideas have been reinterpreted in light of new perspectives on time and social life (Adam 2003) as well as the rise of neoliberal ideology (Glennie and Thrift 1996, Soldatic 2011).

16. Bending the demands of abstract labor in this way is precisely the kind of "soldiering" that Fredrick Taylor characterized as the "greatest evil" and the best justification for scientific management of the labor process (Abbott and Kelly 2005: 89).

17. Zygmunt Bauman (2004: 111) also describes credit/debt relations in contemporary capitalism as a kind of waste relationship, one characteristic of the social formation he calls "liquid modernity."

CHAPTER 3: GOING SHOPPING

1. For a historical account of the rise of consumerism and the moral dilemmas it has posed throughout American history, see Lears (1981, 1994). Associations between consumption and waste predate the rise of mainstream environmentalism. At the end of the nineteenth century, Thorstein Veblen (1899) pilloried the "conspicuous consumption" and wastefulness of the nascent leisure classes. After the post–World War II resurgence of U.S. consumerism, Vance Packard (1960) pursued an analogous attack on the "throwaway society"—a way of life made possible by the planned obsolescence of manufactured commodities and the complementary transformation of the aspirational home into a site of excessive consumption. Daniel Miller (1998, 2012) provides an overview of common critiques of consumption and consumerism as well as their cultural and ideological underpinnings.

2. For more nuanced approaches to the disposal of household goods that avoid moralizing condemnation of waste practices, see Hetherington (2003), Gregson and Crewe (2003), Gregson et al. (2007), and O'Brien (2007).

3. This also goes for waste scholars. Participants in William Rathje's famous garbage project at the University of Arizona had to follow two fundamental rules: do not eat anything and do not take anything home (Randall McGuire, personal communication).

4. For this reason, the practice of searching through mass waste for things of value is generally left to the socially marginal: the underclasses—what Bourgois and Schonberg (2009) call *the lumpen*—as well as artists, outsiders, and activists (see Millar 2008, 2012; Faulk 2012). Giles (2014) has done research on anarchist, food-scavenging cooperatives in various cities throughout the Global North. He argues that they pose a direct challenge to the interests of food retailers, whose profits depend on customers buying new and discriminating on the basis of quality.

5. More discussion of this distinction can be found in Reno (2009). A similar argument motivated Marshall Sahlins (1968) to write his classic critique of popular understandings of those other scavengers in the anthropological literature, hunter-gatherers, whose characterization as impoverished and desperate he contested. Bird-David (1992) offers a more recent reformulation of Sahlins's argument.

6. This is particularly evident today in the electronics and electrical products industry, where newer models are routinely introduced that seem to unceremoniously displace old commodities to which we have formed little or no attachment (see Gabrys 2011).

7. My use of *estrangement* is borrowed from the concept of *ostraenie* in Russian Formalist discourse and the related *Verfremdung* of Brechtian theater (see Mitchell 1974). I am not so much interested in debates about socialist realism in artistic production, but merely want to isolate the distinctiveness of a critical response to capitalist mystification that does not rely upon a separation of the world into reality and appearance (or the really real and the really made up) but takes seriously the role that representation plays in how people relate to things (see Taussig 1989; Pedersen 2008, 2013).

8. In a similar way, Anna Tsing (2005: 51) describes the same problems that confront the commoditization and global circulation of coal.

9. As Fernando Coronil (2001) points out, there are many different forms of capitalist markets, to say nothing of noncapitalist ones, and they do not work equally hard at sustaining the discrete form of the commodities they sell. In this chapter, I attempt to describe the general model that works in the background of the kinds of capitalist exchange most associated with North American mass consumption. This model divides home from market and depicts them in opposing and complementary ways. In other words, this model is about what my informants intimated and stated that markets *should* be like, but all such ideals are only imperfectly realized in the actual world.

10. Representation and matter are thus not ontologically opposed but interwoven, as argued by Keane (2003) utilizing Peircian semiotics.

11. Though valueless in itself, bundling waste could also be seen to represent a modification of the original commodity that further alienates it from its production and provides justification for the higher profit margins of retailers compared with producers. This is how Dudley (2000: 154) describes the relationship between agricultural producers and supermarkets in the Midwest.

12. The financialization of market risks was meant to hedge investors and sellers against the unpredictable hazards of both material circulation and price fluctuation—only to evolve into the derivatives trading held responsible for the recent financial collapse (Cronon 1991).

13. One could usefully compare this to Laura Brown's (2010) ethnography of small shops in India, where an overemphasis on precise accounts (e.g., a receipt) could index a failure to maintain social relationships important for sustaining commerce.

14. There are many aspects to the two-sphere ideology of private and public, and they cannot be reduced to the divide between home and market (see Lopata 1993, Warren 2007). However, the latter distinction is arguably critical for the practice of mass consumption, and I focus on the role of disposability in its sociomaterial constitution for this reason. Adherents to this version of the two-sphere ideology might not claim that home and market are separate today, but they are likely to insist that there was once a time when they were and, more to the point, that they ought to have been, as does Lasch (1995).

15. Annette Weiner (1992) and Maurice Godelier (1999) argue that the sacred–inalienable exist in dynamic tension with the profane–alienable and that practices of giving, keeping, and returning evoke and manipulate this tension for different ends. According to Bloch and Parry (1989), modern capitalist societies tend to invert the ordinary relationship between exchange and possession found nearly everywhere else.

16. The tendency to see these constitutive relations as a matter of labor in and wages out is partly accounted for by the fact, recognized by Marx, that certain fetishized forms (land, labor, capital, and their spectral doubles ground-rent, wages, and profits) suggest themselves as relatable sources of value for interpreting capitalist political economy (see Coronil 1997).

17. Miller (1998) describes the gendered significance of shopping among London consumers and compares it to a form of sacrifice, following Georges Bataille's approach to general economy.

18. Hawkins (2013) depicts food packaging as just such a market device, incorporated to facilitate consumption and calculation. Scholars writing in the material sociology of finance attempt to account for the emergence of the distinctive interpersonal character of markets by examining the various technicalities that cultivate semiotic and behavioral outcomes appropriate to this domain (see Callon 1998, Muniesa et al. 2007, MacKenzie 2009). Though they typically do not, one could also speak of home devices—those technicalities of the private sphere that help cultivate a distinct repertoire of affects and ethics.

19. Michael Arnzen (2010) characterizes collection as a drive to accumulate, proper to capitalist society, whereby collectors seek encounters with the uncanny to restore some sense of what they have lost and may yet lose.

20. As Graeber (2007: 75–8) argues, consumption is about more than the egoistic destruction of things imagined by ideologies of consumerism, because using and reusing things helps to create and refashion social relationships.

21. As Susan Strasser (1999: 10) writes about the domestic stewardship of objects in U.S. households at the end of the twentieth century, "Fixing and finding uses for worn and broken articles entail a consciousness about materials and objects"; not only does reuse rely upon knowledge of the processes by which a thing is produced, but in some cases it may involve "even more creativity than original production" (see also Strathern 2004: 116).

22. This gendered distinction between work and labor is further elaborated in Arendt (1958).

23. Dant (2005) describes car care in some detail, as a social and material practice. Mellström (2004) characterizes tinkering with technology, in general, as an activity that is typically gendered as a form of masculine sociability in Euro-American contexts and beyond.

24. As Horne et al. (2011) argue, homes are filled with spaces like closets, attics, and basements that are used for storage, where disused possessions, spare parts, and ambiguously valued materials are set aside, perhaps one day to be restored or divested. Hetherington (2003) explains the complexity of putting things in

abeyance prior to disposal and removing them later. Given the lingering presence of objects in places where they were once stored, for example, sites like garages often contain the remnants of tinkering, in the form of odors, stains, and spare components.

25. The first uses of "fetishism" by European elites were meant to characterize a primitive desire for spontaneous encounters with worldly things—a racist, Enlightenment-era critique of the supposedly irrational, non-Western other. This is precisely what Marx sought to turn on its head by accusing capitalists of the very same thing (Pietz 1985, 1987; Stallybrass 1999).

CHAPTER 4: WASTELAND HISTORICITY

1. The idea that archaeological findings are not transparent records of the past but, rather, are inevitability mediated by formation processes was a central axiom of behavioral archaeology (Schiffer 1972, 1987; Reid et al. 1975). This perspective also helped legitimize the archaeological investigation of the material remnants of contemporary life, including landfills (see Buchli and Lucas 2001, Shanks et al. 2004, Dawdy 2006, DeSilvey 2006, Harrison and Schofield 2010).

2. While a case could be made that there is less room for landfills in the crowded cities of the Northeast today, there is abundant dumping space in North America in general (Rathje and Murphy 2001: 106–9). It is this geographical absence that makes possible a few very big landfills and makes them more likely in some places rather than others.

3. For an anthropological account of the politics of historicity, see Michel-Rolph Trouillot (1995).

4. An early and influential representation of America both as a new Eden and as going to waste in native hands can be found in John Locke's seminal essay "Of Property" in *Two Treatises of Government*. Regarding the myth of the frontier, see Smith (1950), Kolodny (1975), and Slotkin (1985) on the United States and Furniss (2011) on Canada.

5. Uneven development is used by Marxist geographers to characterize the seesaw flows of capital that shape perpetual patterns of investment and divestiture in country and city (Smith 1990, 2003; Darling 2005). Gidwani and Reddy (2011), Yates (2011), and Gidwani (2013) argue, in a similar way, that waste is a necessary complement to capitalist development, whether through the privatization of a "tragically" misused commons or in the excess that comes with obsolescence, overproduction, and disposable labor.

6. See Rogers (2005: 197–8) and Harvey (1996: 366–9). The environmental-justice literature demonstrates how firms and polities tend to locate polluting sites in the vicinity of marginalized populations. See chapter 5 for more discussion.

7. On the ambiguous place of hillbillies in the contested poetics of whiteness in America, see Hartigan (1999). I discuss local practices of racialization and interpretive repertoires of social difference in chapter 5.

8. In an ethnography of rural life amid the ruins and hills of West Virginia, Kathleen Stewart (1996: 92–3) contrasts official History—born of assembled, analyzed, and narrativized facts—with interactive processes of remembering and "unforgetting" associated with roaming through familiar landscapes.

9. At the time, St. Louis was the second-largest city in the Midwest and a major hub of western commerce (Teaford 1993: 49). The proposed line provided the most straightforward path between the St. Louis main line and the depot in downtown Detroit.

10. For a comprehensive history of Michigan's uneven development and its relationship to regional and national economic transformations in the nineteenth century, see Lewis (2002).

11. On the shifting politics of whiteness in the colonization of the Old Northwest, see White (1991). The problem of "poor whites" would continue to serve as a matter of state policy and racializing practice into the twentieth century, leading settler states to build and share expertise and institutions (Stoler 2006: 50–4).

12. On rethinking western settlement of North America as a process of internal colonization, see Williams (1958), Cayton and Onuf (1990), Gordon (1999), Watts (2001), and Nye (2003). Ann Stoler (2006, 2007, 2013) has argued that imperial formations are not fixed or stable forms of rule but routinely involve shifting and exceptional arrangements of power and ruin, which challenge the divides typically assumed by official imperial cartography and historiography (e.g., between core and periphery, white and nonwhite).

13. Jefferson proposed a Lockean design for national uniformity through labor, land, and government: "When once you have property, you will want laws and magistrates to protect your property . . . you will unite yourselves with us . . . and form one people with us, and we shall all be Americans" (White 1991: 474; see also Hannah 2000: 118)

14. Davies et al. (2003) provide an overview of the problems faced by small towns in the American Midwest in the twenty-first century.

15. This did not come without a cost, however. As I discuss in chapter 5, some of the activists hold County Services accountable for the mysterious deaths of two of their friends who helped oppose the landfill in the 1980s. Rumors circulate, for example, that Bruno used violent Mafia tactics to protect his company's business interests.

16. For a discussion of the past and future in relation to the absent presence of the dead, see Kohn (2013: 23–4, 191–220).

CHAPTER 5: GHOSTLY AND FLESHLY LINES

1. The transboundary shipment of discards has continued to grow as toxics, recyclates, and waste electrical and electronic equipment become more expensive to manage domestically and as the definitions of *hazardous* and *recycling* become more flexible (see Clapp 2001, Gabrys 2011, Alexander and Reno 2012, Lepawsky 2014).

2. By comparison, most states import less than a million tons of waste per year. The leading waste importer in the United States for the past several decades has been Pennsylvania at 10.5 million tons in 2005, or more than 20 percent of the nation's waste imports (Simmons et al. 2006: 32).

3. For a longer discussion of animal waste as related to identity and difference, see Serres (2010) and Reno (2014).

4. My distinction between ghostly lines and the traces materially inscribed in the landscape comes from Ingold (2007: 47–9). On one hand, national borders are ghostly in that they are always subject to continual processes of *rebordering,* or transformations in state policy, public perception, and economic circulation. In the case of North America, this has involved ongoing disputes over security, fair trade, smuggling, immigration, and environmental pollution (Andreas and Biersteker 2003). On the other hand, the trauma of border crossing leaves behind ghostly figures; Khosravi (2011) wrote an autoethnography of the violent "bordering" of people who cross national lines, the metaphorical and literal ghostliness of whom is Langford's (2013) focus.

5. As abstractions, both racial and national lines oversimplify and obfuscate the real conditions of the world—national borders abstract from the social and political realities of the borderlands that preexist and prefigure them (Donnan and Wilson 2010: 8), and racialization abstracts from the nuances of actual human interaction and embodiment (Urciuoli 1996; Herzfeld 1997; Hill 1998, 2009).

6. Referring to waste circulation in terms of geoeconomic distinctions between rich and poor countries is now an established rhetorical device in international waste discourse, but one that unintentionally and implicitly *depoliticizes* trade between rich countries or between poor countries (O'Neill 2000, Lepawsky 2014).

7. Briggs and Briggs (2006: 10) illustrate how racial economy influences normative expectations about health and illness that represent unmarked members of the population as "sanitary citizens" while others are cast as ignorant or dirty "unsanitary subjects."

8. Sugrue (1996) offers a detailed historical account of the marginalization of African-American workers in Detroit and its relationship to the city's twentieth-century decline. Zimring (2004, 2009) investigates the history of scrap industries and discrimination in the United States. Reno (2015) offers an overview of the relationship between filthy work and hierarchy in recent ethnographic literature and anthropological theory.

9. Whether one considers the nuclear colonialism of radioactive-waste depots and weapons testing or the more recent tar-sands operation in Alberta, North American indigenous communities tend to be disproportionately affected by the destructive remains of settler societies (see Thorpe 1996, Masco 2006, Mascarenhas 2007, Johnston and Barker 2008, Endres 2009, Kobayashi et al. 2011).

10. On environmental-justice movements and scholarship, especially in North America, see Bullard (1990, 2005), Bryant and Mohai (1992), Pellow (2007), and Carmin and Agyeman (2011).

11. On First Nation resistance to mega-projects, see Nadasdy (2003: 53–5). On effects of hydroelectric development on the James Bay Cree, see Niezen (2003: 151–2).

12. After their successful opposition to Toronto's waste scheme, Timiskaming First Nation became embroiled in a new dispute following the discovery of indigenous remains on the site of the historic Fort-Témiscamingue/Obadjiwan (Neveu 2010: 246–51).

13. The amount of Canadian waste dumped began to decline in 2007, partially as a result of political and public opposition to the practice but also because of rising transportation costs and improved recycling programs in Canada.

14. Wendy Brown (1995) calls this form of identification within political liberalism "wounded attachment" because it achieves public recognition through investment in subjection and weakness. Brown argues that this *ressentiment* is not wholly negative but is rather the general product of a tension between freedom and equality within political liberalism.

15. The idea of such a "material public" of affected people is developed in Marres (2012).

16. On public forums, see Callon et al. (2002). The idea of public forums is applicable to the governance of landfills insofar as the object of reflexive processes like the MDEQ hearing is a transformation of the *sociotechnical capacity* of landfill services, or "human competencies and material devices that have been designed and arranged in a way in which they can be mobilized in order to achieve desired results" (Callon et al. 2002: 208). Since available air space is a part of such capacity, expansion permits are a part of the qualification of waste as a circulated bad.

17. See Little (2014) on the internal tensions that threaten grassroots environmental movements.

18. For accounts of the U.S.–Canadian borderlands and border relations, see McDougall and Philips (2012), McGrady (2006), and Ramirez (2001).

19. Roediger (1999) has described the real privilege and imagined superiority of whites as a "psychological wage" that is provided to white lower classes in the United States. Hartigan (1999) offers a detailed analysis of the poetics of racial difference among whites in the Detroit area specifically.

20. To the extent that whites are marked as "ethnic" in the contemporary United States, it is most often framed as a choice made by those who have the privilege to embrace their ancestry or not (Di Leonardo 1998). According to Graeber (2006), ignorance about people of lower status often plays an important role in relations of hierarchy.

21. According to an image popularized by transnational textual forms, including books, newspapers, and websites, the Sikh subject is the man abroad, working hard in order to perpetuate the Sikh *panth* (or community) and the *quam* (or nation). The "total body" of the *amritdhari* is the iconic embodiment of the global Sikh community and the fantasy of a coming national homeland (Axel 2001: 150). The total body of the *amritdhari* is made an object of public surveillance (Axel 2001: 42; see

also Fox 1985). In other words, the bodies of Sikh men are meant to make these otherwise ghostly identifications and aspirations concrete.

22. It may be that talking to a young, white American male led Bula to downplay the racism he encountered while on the road or at Four Corners. In addition to the public resentment they face for employment in the waste trade, Sikh drivers may be alienated by linguistic and cultural differences during the course of their journeys, which force them to leave the ethnic enclaves of urban Toronto to which they are more accustomed. Any yet, like Bula, those Sikh drivers I spoke with claimed not to be particularly troubled by their misrepresentation and occasional mistreatment while delivering to Four Corners. Perhaps, given the trauma of racist riots and persecution against Sikhs in India from 1984 onward, prejudice and misunderstanding while hauling waste might seem rather mild by comparison.

23. A literal interpretation is "a reading (whether actually literal or not) that stresses what is taken to be the standard meaning of the sentence—its propositional content—and suppresses all other possibilities" (Goffman 1981: 56).

24. Asif Agha argues that when people make use of linguistic registers, they tend to index certain voices, or social personae, in the process: "[E]ncounters with registers are . . . encounters in which individuals establish forms of footing and alignment with voices indexed by speech and thus with social types of persons, real or imagined, whose voices they take them to be" (2005: 38).

25. Andreas (2003), Hristoulas (2003), and Miller (2006) all present accounts that challenge a simplistic view of Canadian–U.S. border relations as binational cooperation.

26. State institutions also fixate on people of actual and mistaken Arab identity as possible threats to security. When one of the laborers at Four Corners was interrogated by FBI agents as part of a narcotics investigation, for example, they accused him of using Canadian truck drivers as suppliers for the ecstasy pills he was selling during his time off from the landfill. His actual supplier was a man from the metro area, but he was happy to let them continue believing that the waste trade was a front for international drug smuggling. Cases of mistaken identity can be used against state surveillance as easily as they can be deployed in its favor.

27. In this, the people of southeastern Michigan are no different from their Timiskaming counterparts, whose successful self-determination arguably cannot be separated from a degree of submission to the capitalist development of indigenous-held lands (see Slowey 2008, Eisenberg et al. 2014).

CONCLUSIONS

1. As filth and pollution became associated with peasant life in Sweden, becoming bourgeoisie meant embracing separation from transience. See Frykman and Lofgren (1987) for a cultural and historical analysis.

2. Though even here, Sweden seems to exceed expectations. Their new nuclear storage facilities have reportedly been met with high rates of local acceptance and an apparent lack of NIMBYism (see Sjöberg 2004).

3. See note 5 to the Introduction.

4. See Connett (2013: 37–9) on interim landfilling and Johansson et al. (2012) on the promises of, and obstacles to, landfill excavation in Sweden as a way to reorient the dump regime toward more reuse.

5. More broadly, this would represent a shift toward a steady-state or degrowth economy, one that recognizes the bubble-like, unsustainable foundations of capitalist accumulation of new and everlasting forms (see Foster 2002, Latouche 2009).

6. Žižek (2010: 35) and Gabrys (2011: 150) both discuss problems with the idea of zero waste.

7. Rotman (1987) offers an insightful and original semiotic analysis and genealogical critique of the significance of zero as a number and representational practice.

8. Both Smith (1988) and Graeber (2011b) analyze the relationship between the desire for money and the pursuit of higher values—which have been historically opposed but, thereby, also bound to one another.

9. Huntington and Metcalf (1979) wrote a classic account of death rituals in North America. Sørensen (2009) and Driskill (2012) offer more recent analyses of the material significance of headstones in funerary practice.

10. Such mistreatment of the dead is especially troubling when one views the washing or purification of a dead body as a necessary part of mourning and funerary ritual (see Connor 1995, Abramovich 2005).

11. On the power of corpses in the regeneration of life, see Molleson (1981) and Bloch and Parry (1982). Rahtz (1981) and Strathern (1981) discuss the importance of grave goods for the living.

12. In her ethnography, Henrietta Moore (1986: 103–5) makes clear that these choices of burial site are never absolute, however, but only practical rules of thumb.

13. Throughout this book, I have referred to "North Americans," an admittedly clumsy and problematic toponym. I have done so because the waste practices I have been describing are shared by, and unite, people in the United States and Canada, not because they represent a bounded territorial whole or culture (see note 2 to the Introduction). There are, as I will now discuss, people living within the ghostly borders of this artificial subcontinent who engage with and imagine waste in very different ways.

14. My analysis of Pueblo cosmology relies upon the work of Ellis (1952, 1966), Levi-Strauss (1955), Parmentier (1979), Parsons (1980), Silko (1986), Walker and Lucero (2000), and Cameron (2002). It bears mentioning that interpretations of Pueblo cosmology are severely constrained by secrecy and the limited access granted to outsiders.

15. For an analysis of the co-optation of the environmentalist cause by corporate interests in the United States, see Rogers (2005: 145).

16. Pueblo doings, as Severin Fowles (2013) has argued, are not easily classified as premodern or modern, sacred or profane, but are a form of cosmological practice that refuses these illusory divides—divides that moderns seem to insist on living by.

17. Bauman goes on to argue that "there is little such experience left now—in the quicksands of protean, kaleidoscopic sights" (2004: 105). His provocative discussion of everyday encounters with what he calls "eternity," and their subsequent loss in modern life, calls to mind Roy Rappaport's (1999) more elaborate argument about the importance of communicative ritual acts in human experience and their avowal of "eternal verities" about the cosmos. Despite many similarities, where Bauman emphasizes the opposition between eternity and the relative liquidity of modern existence, for Rappaport permanence and impermanence are not simply in tension, but are more dynamically related as general conditions of existence (see Lambek 2001).

BIBLIOGRAPHY

Abbott, Keith and Peter Kelly. 2005. Conceptualizing Industrial Relations in the 'Risk Society.' *Labour & Industry* 16(1): 85–102.

Abramovich, Henry. 2005. Where Are the Dead? Bad Death, the Missing, and the Inability to Mourn. In S. Heilman, ed. *Death, Bereavement, and Mourning.* New Brunswick, NJ: Transaction. pp. 53–68.

Adam, Barbara. 1998. *Timescapes of Modernity: The Environment and Invisible Hazards.* London: Routledge.

———. 2003. When Time Is Money: Contested Rationalities of Time in the Theory and Practice of Work. *Theoria: A Journal of Social and Political Theory* 102: 94–125.

Agha, Asif. 2005. Voice, Footing, Enregisterment. *Journal of Linguistic Anthropology* 15(1): 38–59.

Alexander, Catherine and Joshua Reno, eds. 2012. *Economies of Recycling: Global Transformations of Materials, Values and Social Relations.* London: Zed Books.

———. 2014. From Biopower to Energopolitics in England's Modern Waste Technology. *Anthropological Quarterly* 87(2): 335–58.

Alexander, Judd. 1993. *In Defense of Garbage.* Westport, CT: Praeger.

Andreas, Peter. 2003. A Tale of Two Borders: The U.S.–Canada and U.S.–Mexico Lines after 9–11. In P. Andreas and T. J. Biersteker, eds. *The Rebordering of North America: Integration and Exclusion in a New Security Context.* New York: Routledge. pp. 1–23.

Andreas, Peter and Thomas J. Biersteker, eds. 2003. *The Rebordering of North America: Integration and Exclusion in a New Security Context.* New York: Routledge.

Arendt, Hannah. 1958. *The Human Condition.* Chicago: University of Chicago Press.

Ariew, André and Richard Lewontin. 2004. The Confusion of Fitness. *British Journal for the Philosophy of Science* 55(2): 347–63.

Armstrong, David. 2002. *A New History of Identity: A Sociology of Medical Knowledge.* New York: Palgrave Macmillan.

Arnzen, Michael. 2010. The Uncanny Impulse to Collect. http://gorelets.com /uncanny/theory/the-uncanny-impulse-to-collect/.

Ashforth, Blake E. and Glen E. Kreiner. 1999. "How Can You Do It?": Dirty Work and the Challenge of Constructing a Positive Identity. *Academy of Management Review* 24 (3): 413–34.

———. 2014. Dirty Work and Dirtier Work: Differences in Countering Physical, Social, and Moral Stigma. *Management and Organization Review* 10(1): 81–108.

Axel, Brian K. 2001. *The Nation's Tortured Body: Violence, Representation, and the Formation of the Sikh Diaspora*. Durham, NC: Duke University Press.

Bakhtin, Mikhail. 1965. *Rabelais and His World*. Bloomington: Indiana University Press.

Barillas, William. 1989. Michigan's Pioneers and the Destruction of the Hardwood Forest. *Michigan Historical Review* 15(2): 1–22.

Barthes, Roland. 1989. *The Rustle of Language*. New York: Farrar, Straus and Giroux.

Bateson, Gregory. 1972. *Steps to an Ecology of Mind*. Chicago: University of Chicago Press.

Bauman, Zygmunt. 2004. *Wasted Lives: Modernity and Its Outcasts*. Cambridge, UK: Polity Press.

Beck, Ulrich. 1992. *Risk Society: Towards a New Modernity*. New York: Sage.

Bellah, Robert, Richard Madsen, William Sullivan, Ann Swidler, and Steven Tipton. 1986. *Habits of the Heart: Individualism and Commitment in American Life*. New York: Harper and Row.

Bennett, Jane. 2009. *Vibrant Matter: Toward a Political Ecology of Things*. Durham, NC: Duke University Press.

Bettie, Julie. 2003. *Women without Class: Girls, Race, and Identity*. Berkeley: University of California Press.

Bird-David, Nurit. 1992. Beyond "The Original Affluent Society." *Current Anthropology* 33(1): 25–47.

Blackburn, Robin. 2006. *Age Shock: How Finance Is Failing Us*. London: Verso.

Bloch, Maurice and Jonathan Parry, eds. 1982. *Death and the Regeneration of Life*. Cambridge, UK: University of Cambridge Press.

———, eds. 1989. *Money and the Morality of Exchange*. Cambridge, UK: University of Cambridge Press.

Bolinger, Dwight. 1973. Ambient *It* Is Meaningful Too. *Journal of Linguistics* 9: 261–70.

Bourgois, Philippe and Jeffrey Schonberg. 2009. *Righteous Dopefiend*. Berkeley: University of California Press.

Briggs, Charles and Clara S. Briggs. 2006. *Stories in a Time of Cholera: Racial Profiling during a Medical Nightmare*. Berkeley: University of California Press.

Brown, Kathleen. 2009. *Foul Bodies: Cleanliness in Early America*. New Haven, CT: Yale University Press.

Brown, Laura. 2010. *Tipping Scales with Talk: Conversation, Commerce, and Obligation on the Edge of Thanjavur, India*. PhD dissertation, University of Michigan, Ann Arbor.

Brown, Wendy. 1995. *States of Injury: Power and Freedom in Late Modernity*. Princeton, NJ: Princeton University Press.

Bryant, Bunyan and Paul Mohai, eds. 1992. *Race and the Incidence of Environmental Hazards: A Time for Discourse*. Boulder, CO: Westview Press.

Buchli, Victor and Gavin Lucas, eds. 2001. *Archaeology of the Contemporary Past*. London: Routledge.

Buck-Morss, Susan. 1992. Aesthetics and Anesthetics: Walter Benjamin's Artwork Essay Reconsidered. *October* 62: 3–41.

Bullard, Robert. 1990. *Dumping in Dixie: Race, Class, and Environmental Quality*. Boulder, CO: Westview Press.

———. 2005. *The Quest for Environmental Justice: Human Rights and the Politics of Pollution*. San Francisco: Sierra Club Books.

Burawoy, Michael. 1979. *Manufacturing Consent: Changes in the Labor Process under Monopoly Capitalism*. Chicago: University of Chicago Press.

Burrell, Jenna. 2012. *Invisible Users: Youth in the Internet Cafes of Urban Ghana*. Cambridge, MA: MIT Press.

Bynum, Caroline. 1991. *Fragmentation and Redemption: Essays on Gender and the Human Body in Medieval Religion*. New York: Zone Books.

Callon, Michel. 1998. Introduction: The Embeddedness of Economic Markets in Economics. In M. Callon, ed. *The Laws of Markets*. Cambridge, MA: Blackwell. pp. 1–57.

Callon, Michel, Cecile Meadel, and Vololona Rabeharisoa. 2002. The Economy of Qualities. *Economy and Society* 31(2): 194–217.

Cameron, Catherine. 2002. Sacred Earthen Architecture in the Northern Southwest: The Bluff Great House Berm. *American Antiquity* 67(4): 677–95.

Carmin, JoAnn and Julian Agyeman, eds. 2011. *Environmental Inequalities beyond Borders: Local Perspectives on Global Injustices*. Cambridge, MA: MIT Press.

Cayton, Andrew R.L. and Peter S. Onuf. 1990. *The Midwest and the Nation: Rethinking the History of an American Region*. Bloomington: Indiana University Press.

Chakrabarty, Dipesh. 1992. Of Garbage, Modernity and the Citizen's Gaze. *Economic and Political Weekly* 27(10–11): 7–14.

Chiang, Connie Y. 2008. The Nose Knows: The Sense of Smell in American History. *Journal of American History* 95(2): 405–16.

Clapp, Jennifer. 2001. *Toxic Exports: The Transfer of Hazardous Wastes from Rich to Poor Countries*. Ithaca, NY: Cornell University Press.

Collier, Roger. 2011. The Ethics of Reusing Single-Use Devices. *Canadian Medical Association Journal* 183(11): 1245.

Connett, Paul. 2013. *The Zero Waste Solution: Untrashing the Planet One Community at a Time*. White River Junction, VT: Chelsea Green.

Connolly, William. 2013. *The Fragility of Things: Self-Organizing Processes, Neoliberal Fantasies and Democratic Activism*. Durham, NC: Duke University Press.

Connor, Linda H. 1995. The Action of the Body on Society: Washing a Corpse in Bali. *Journal of the Royal Anthropological Institute* 1(3): 537–59.

Corbin, Alain. 1986. *The Foul and the Fragrant: Odor and the French Social Imagination.* Cambridge, MA: Harvard University Press.

Coronil, Fernando. 1997. *The Magical State: Nature, Money, and Modernity in Venenzeula.* Chicago: University of Chicago Press.

———. 2001. Smelling Like a Market. *American Historical Review* 106(1): 119–29.

Cronon, William. 1991. *Nature's Metropolis: Chicago and the Great West.* New York: W.W. Norton.

Crooks, Harold. 1983. *Dirty Business: The Inside Story of the New Garbage Agglomerates.* Toronto: Lorimer.

———. 1993. *Giants of Garbage: The Rise of the Global Waste Industry and the Politics of Pollution.* Toronto: Lorimer.

Curtis, Valerie, Mícheál de Barra, and Robert Aunger. 2011. Disgust as an Adaptive System for Disease Avoidance Behaviour. *Philosophical Transactions of the Royal Society B* 366: 389–401.

Dant, Tim. 2005. *Materiality and Society.* Maidenhead, UK: Open University Press.

Darling, Eliza. 2005. The City in the Country: Wilderness Gentrification and the Rent Gap. *Environment and Planning A* 37: 1015–32.

Darwin, Charles. 2005 [1872]. *The Expression of the Emotions in Man and Animals.* London: HarperCollins.

Davies, Richard O., David R. Pichaske, and Anthony Amato, eds. 2003. *A Place Called Home: Writings on the Midwestern Small Town.* St. Paul, MN: Borealis Books.

Dawdy, Shannon L. 2006. The Taphonomy of Disaster and the (Re)Formation of New Orleans. *American Anthropologist* 108(4): 719–30.

———. 2010. Clockpunk Anthropology and the Ruins of Modernity. *Current Anthropology* 51(6): 761–93.

Deacon, Terrence. 2012. *Incomplete Nature: How Mind Emerged from Matter.* New York: W.W. Norton.

Deleuze, Gille. 1994. *Difference and Repetition.* New York: Columbia University Press.

Derrida, Jacques. 1976. *Of Grammatology.* Baltimore: Johns Hopkins University Press.

DeSilvey, Caitlin. 2006. Observed Decay: Telling Stories with Mutable Things. *Journal of Material Culture* 11(3): 318–38.

Di Leonardo, Micaela. 1998. *Exotics at Home: Anthropologies, Others, and American Modernity.* Chicago: University of Chicago Press.

Donnan, Hastings and Thomas Wilson. 2010. *Borderlands: Ethnographic Approaches to Security, Power, and Identity.* Lanham, MD: University Press of America.

Douglas, Mary. 1984 [1966]. *Purity and Danger: An Analysis of Concepts of Pollution and Taboo.* New York: Routledge.

Driskill, Nathan. 2012. Distinction in Death: An Analysis of Individual and Brand Consumption in Contemporary American Funeral Practices. PhD dissertation, University of Missouri, Kansas City.

Dudley, Kathryn. 1994. *The End of the Line: Lost Jobs, New Lives in Postindustrial America*. Chicago: University of Chicago Press.

———. 2000. *Debt and Dispossession: Farm Loss in America's Heartland*. Chicago: University of Chicago Press.

Eisenberg, Avigail, Jeremy Webber, Glen Coulthard, and Andrée Boiselle. 2014. *Recognition versus Self-Determination: Dilemmas of Emancipatory Politics*. Vancouver: University of British Columbia Press.

Elias, Norbert. 1994 [1969]. *The Civilizing Process, vol. 1: The History of Manners*. Oxford, UK: Blackwell.

Ellis, Florence H. 1952. Jemez Kiva Magic and Its Relation to Features of Prehistoric Kivas. *Southwestern Journal of Anthropology* 8(2): 147–63.

———. 1966. The Immediate History of Zia Pueblo as Derived from Excavation in Refuse Deposits. *American Antiquity* 31(6): 806–11.

Elson, Diane. 1979. The Value Theory of Labour. In D. Elson, ed. *Value: The Representation of Labour in Capitalism*. London: CSE Books. pp. 115–80.

Endres, Danielle. 2009. The Rhetoric of Nuclear Colonialism: Rhetorical Exclusion of American Indian Arguments in the Yucca Mountain Nuclear Waste Siting Decision. *Communication and Critical/Cultural Studies* 6(1): 39–60.

Faulk, Karen. 2012. Stitching Curtains, Grinding Plastic: Social and Material Transformation in Buenos Aires. In C. Alexander and J. Reno, eds. *Economies of Recycling: Global Transformations of Materials, Values and Social Relations*. London: Zed Books. pp. 143–63.

Favret-Saada, Jeanne. 2015. *The Anti-Witch*. Chicago: Hau Books.

Feld, Steven. 1996. Waterfalls of Song: An Acoustemology of Place Resounding in Bosavi, Papua New Guinea. In K. Basso and S. Feld, eds. *Senses of Place*. Santa Fe, NM: School of American Research Press. pp. 53–90.

Ferguson, James. 2006. *Global Shadows: Africa in the Neoliberal World Order*. Durham, NC: Duke University Press.

Fessler, Daniel and Kevin Haley. 2006. Guarding the Perimeter: The Outside–Inside Dichotomy in Disgust and Bodily Experience. *Cognition & Emotion* 20(1): 3–19.

Foster, John B. 2002. *Ecology against Capitalism*. New York: Monthly Review Press.

Foucault, Michel. 1997. *Society Must Be Defended*. New York: Macmillan.

Fowles, Severin. 2010. People without Things. In M. Bille, T. Sørensen, and F. Hastrup, eds. *An Anthropology of Absence: Materializations of Transcendence and Loss*. Springer: New York. pp. 23–41.

———. 2013. *An Archaeology of Doings: Secularism and the Study of Pueblo Religion*. Santa Fe, NM: School for Advanced Research Press.

Fox, Richard G. 1985. *Lions of the Punjab: Culture in the Making*. Berkeley: University of California Press.

Frank, Dana. 1999. *Buy American: The Untold Story of Economic Nationalism.* Boston: Beacon Press.

Frederick, James and Nancy Lessin. 2000. Blame the Worker: The Rise of Behavioural-Based Safety Programs. *Multinational Monitor* 21(11): 10–7.

Frykman, Jonas and Orvar Lofgren. 1987. *Culture Builders: A Historical Anthropology of Middle-Class Life.* New Brunswick, NJ: Rutgers University Press.

Furniss, Elizabeth. 2011. *The Burden of History: Colonialism and the Frontier Myth in a Rural Canadian Community.* Vancouver: University of British Columbia Press.

Gabrys, Jennifer. 2011. *Digital Rubbish: A Natural History of Electronics.* Ann Arbor: University of Michigan Press.

Gatens, Moira. 1996. *Imaginary Bodies: Ethics, Power and Corporeality.* New York: Routledge.

Gelfand, Alexander. 2007. Green Dream: Shrink-Wrapped Trash. www.wired.com /science/discoveries/news/2007/03/72913.

Gibson-Graham, J. K. 1996. *The End of Capitalism (As We Knew It): A Feminist Critique of Political Economy.* Minneapolis: University of Minnesota Press.

Gidwani, Vinay. 2013. Six Theses on Waste, Value, and Commons. *Social & Cultural Geography* 14(7): 773–83.

Gidwani, Vinay and Rajyashree Reddy. 2011. The Afterlives of "Waste": Notes from India for a Minor History of Capitalist Surplus. *Antipode* 43(5): 1625–58.

Giles, David. 2014. The Anatomy of a Dumpster: Abject Capital and the Looking Glass of Value. *Social Text* 32 (1 118): 93–113.

Gille, Zsuza. 2007. *From the Cult of Waste to the Trash Heap of History: The Politics of Waste in Socialist and Postsocialist Hungary.* Bloomington: Indiana University Press.

Glennie, Paul and Nigel Thrift. 1996. Reworking E. P. Thompson's 'Time, Work-Discipline and Industrial Capitalism.' *Time & Society* 5(3): 272–99.

Godelier, Maurice. 1999. *The Enigma of the Gift.* Chicago: University of Chicago Press.

Goffman, Erving. 1963. *Stigma: Notes on the Management of Spoiled Identity.* New York: Simon & Schuster.

———. 1981. *Forms of Talk.* Philadelphia: University of Pennsylvania Press.

Gordon, Linda. 1999. *The Great Arizona Orphan Abduction.* Cambridge, MA: Harvard University Press.

Graeber, David. 2001. *Toward an Anthropological Theory of Value: The False Coin of Our Own Dreams.* New York: Palgrave.

———. 2006. Beyond Power/Knowledge: An Exploration of the Relationship of Power, Ignorance, and Stupidity. Malinowski Memorial Lecture, London School of Economics and Political Science.

———. 2007. *Possibilities: Essays on Hierarchy, Rebellion, and Desire.* Oakland, CA: AK Press.

———. 2011a. *Debt: The First 5000 Years.* New York: Melville House.

————. 2011b. The Divine Kingship of the Shilluk: On Violence, Utopia, and the Human Condition, or, Elements for an Archaeology of Sovereignty. *Hau: Journal of Ethnographic Theory* 1(1): 1–62.

————. 2013. It Is Value That Brings the Universe into Being. *Hau: Journal of Ethnographic Theory* 3(2): 219–43.

Gregson, Nicky and Mike Crang. 2010. Materiality and Waste: Inorganic Vitality in a Networked World. *Environment and Planning A* 42: 1026–32.

Gregson, Nicky and Louise Crewe. 2003. *Second-Hand Cultures*. London: Bloomsbury Academic.

Gregson, Nicky, Alan Metcalfe, and Louise Crewe. 2007. Moving Things Along: The Conduits and Practices of Divestment in Consumption. *Transactions of the Institute of British Geographers* 32(2): 187–200.

Griffiths, Paul E. 1997. *What Emotions Really Are: The Problem of Psychological Categories*. Chicago: University of Chicago Press.

Halvorson, Britt. 2012. 'No Junk for Jesus': Redemptive Economies and Value Conversions in Lutheran Medical Aid. In C. Alexander and J. Reno, eds. *Economies of Recycling: Global Transformations of Materials, Values and Social Relations*. London: Zed Books. pp. 207–33.

Hannah, Matthew G. 2000. *Governmentality and the Mastery of Territory in Nineteenth Century America*. Cambridge, UK: Cambridge University Press.

Harman, Graham. 2005. *Guerilla Metaphysics: Phenomenology and the Carpentry of Things*. Peru, IL: Open Court.

————. 2009. *Prince of Networks: Bruno Latour and Metaphysics*. Melbourne: re.press.

Harrison, Rodney and John Schofield. 2010. *After Modernity: Archaeological Approaches to the Contemporary Past*. Oxford, UK: Oxford University Press.

Hart, John Fraser. 1975. *The Look of the Land*. Englewood Cliffs, NJ: Prentice-Hall.

Hartigan, John. 1992. Reading Trash: Deliverance and the Poetics of White Trash. *Visual Anthropology Review* 8(2): 8–15.

————. 1997. Unpopular Culture: The Case of "White Trash." *Cultural Studies* 11(2): 316–43.

————. 1999. *Racial Situations: Class Predicaments of Whiteness in Detroit*. Princeton, NJ: Princeton University Press.

Harvey, David. 1996. *Justice, Nature and the Geography of Difference*. London: Wiley-Blackwell.

————. 2005. *A Brief History of Neoliberalism*. Oxford, UK: Oxford University Press.

Hawkins, Gay. 2006. *The Ethics of Waste: How We Relate to Rubbish*. Lanham, MD: Rowman and Littlefield.

————. 2013. The Performativity of Food Packaging: Market Devices, Waste Crisis and Recycling. *Sociological Review* 69(S2): 66–83.

Herzfeld, Michael. 1997. *Cultural Intimacy: Social Poetics in the Nation-State*. New York: Routledge.

Hetherington, Kevin. 2003. Secondhandedness: Consumption, Disposal, and the Absent Presence. *Environment and Planning D: Society and Space* 22(1): 157–73.

Hetrick, William P. and David M. Boje. 1992. Organization and the Body: Post-Fordist Dimensions. *Organizational Change Management* 5(1): 48–57.

Hill, Jane. 1998. Language, Race, and White Public Space. *American Anthropologist* 100(3): 680–9.

———. 2009. *The Everyday Language of White Racism*. London: Wiley-Blackwell.

Hird, Myra. 2012. Knowing Waste: Towards an Inhuman Epistemology. *Social Epistemology* 26(3–4): 453–69.

Holmes, Douglas and George Marcus. 2005. Cultures of Expertise and the Management of Globalization: Toward the Re-Functioning of Ethnography. In A. Ong and S.J. Collier, eds. *Global Assemblages: Technology, Politics and Ethics as Anthropological Problems*. London: Wiley Blackwell. pp. 235–52.

Horne, Ralph, Cecily Maller, and Ruth Lane. 2011. Remaking Home: The Reuse of Goods and Materials in Australian Households. In R. Lane and A. Gorman-Murray, eds. *Material Geographies of Household Sustainability*. Farnham, UK: Ashgate.

Hoy, Suellen. 1996. *Chasing Dirt: The American Pursuit of Cleanliness*. Oxford, UK: Oxford University Press.

Hristoulas, Athanasios. 2003. Trading Places: Canada, Mexico, and North American Security. In P. Andreas and T.J. Biersteker, eds. *The Rebordering of North America: Integration and Exclusion in a New Security Context*. New York: Routledge. pp. 24–45.

Hughes, Everett C. 1951. Mistakes at Work. *Canadian Journal of Economics and Political Science* 17(3): 320–27.

———. 1962. Good People and Dirty Work. *Social Problems* 10(1): 3–11.

Humes, Edward. 2012. *Garbology: Our Dirty Love Affair with Trash*. New York: Penguin.

Huntington, Richard and Peter Metcalf. 1979. *Celebrations of Death: The Anthropology of Mortuary Ritual*. New York: Cambridge University Press.

Iacuone, David. 2005. "Real Men Are Tough Guys": Hegemonic Masculinity and Safety in the Construction Industry. *Journal of Men's Studies* 13(2): 247–66.

Ingold, Tim. 2004. Culture on the Ground: The World Perceived through the Feet. *Journal of Material Culture* 9(3): 315–40.

———. 2007. *Lines: A Brief History*. London: Routledge.

———. 2010. *Bringing Things to Life: Creative Entanglements in a World of Materials*. Working Paper, Department of Anthropology, University of Manchester, UK.

———, ed. 2011. *Redrawing Anthropology: Materials, Movements, Lines*. Farnham, UK: Ashgate.

Ingold, Tim and Gisli Palsson, eds. 2013. *Biosocial Becomings: Integrating Social and Biological Anthropology*. Cambridge, UK: Cambridge University Press.

Jablonksi, Nina. 2006. *Skin: A Natural History*. Berkeley: University of California Press.

Jackson, Deborah D. 2011. Scents of Place: The Dysplacement of a First Nations Community in Canada. *American Anthropologist* 113(4): 606–18.

Jervis, Lori L. 2001. The Pollution of Incontinence and the Dirty Work of Caregiving in a US Nursing Home. *Medical Anthropological Quarterly* 15(1): 84–99.

Johansson, Nils, Joakim Krook, and Mats Eklund. 2012. Transforming Dumps into Gold Mines: Experiences from Swedish Case Studies. *Environmental Innovation and Societal Transitions* 5: 33–48.

Johnston, Barbara R. and Holly M. Barker. 2008. *The Consequential Damages of Nuclear War*. Walnut Creek, CA: Left Coast Press.

Joyce, Patrick. 2003. *The Rule of Freedom: Liberalism and the Modern City*. London: Verso.

Juarrero, Alicia. 1998. Causality as Constraint. In G. Van de Vijver, S. Salthe, and M. Delpos, eds. *Evolutionary Systems*. Dordrecht, The Netherlands: Kluwer Academic. pp. 233–42.

———. 1999. *Dynamics in Action: Intentional Behavior as a Complex System*. Cambridge, MA: MIT Press.

Kahn, Howie. 2011. Destroying Detroit (in Order to Save It). *GQ*. www.gq.com /news-politics/big-issues/201105/detroit-renovation?printable = true.

Keane, Webb. 2003. Semiotics and the Social Analysis of Material Things. *Language and Communication* 23: 409–25.

Kennedy, Greg. 2008. *An Ontology of Trash: The Disposable and Its Problematic Nature*. Albany, NY: SUNY Press.

Kenyon, Amy M. 2004. *Dreaming Suburbia: Detroit and the Production of Postwar Space and Culture*. Detroit: Wayne State University Press.

Khosravi, Shahram. 2011. *"Illegal" Traveller: An Auto-Ethnography of Borders*. New York: Palgrave Macmillan.

Klepeis, Neil, William Nelson, Wayne Ott, John Robinson, Andy Tsang, Paul Switzer, Joseph Behar, Stephen Hern, and William Engelmann. 2001. The National Human Activity Pattern Survey (NHAPS): A Resource for Assessing Exposure to Environmental Pollutants. *Journal of Exposure Analysis and Environmental Epidemiology* 11: 231–52.

Kobayashi, Audrey, Laura Cameron, and Andrew Baldwin. 2011. *Rethinking the Great White North: Race, Nature and Whiteness in Canada*. Vancouver: University of British Columbia Press.

Kohn, Eduardo. 2013. *How Forests Think: Toward an Anthropology Beyond the Human*. Berkeley: University of California Press.

Kolnai, Aurel. 2004. *On Disgust*. Peru, IL: Open Court.

Kolodny, Annette. 1975. *The Lay of the Land: Metaphor as Experience and History in American Life and Letters*. Chapel Hill: University of North Carolina Press.

Kristeva, Julia. 1982. *Powers of Horror: An Essay on Abjection*. New York: Columbia University Press.

Lambek, Michael. 2001. Rappaport on Religion: A Social Anthropological Reading. In R. A. Rappaport, E. Messer, and M. Lambek, eds. *Ecology and the Sacred: Engaging the Anthropology of Roy A. Rappaport*. Ann Arbor: University of Michigan Press. pp. 244–74.

Langford, Jean. 2013. *Consoling Ghosts: Stories of Medicine and Mourning from Southeast Asians in Exile*. Minneapolis: University of Minnesota Press.

Laporte, Dominic. 2002. *History of Shit*. Cambridge, MA: MIT Press.

Lasch, Christopher. 1995. *Haven in a Heartless World: The Family Besieged*. New York: Basic Books.

Latouche, Serge. 2009. *Farewell to Growth*. Cambridge, UK: Polity.

Latour, Bruno. 1993. *We Have Never Been Modern*. Cambridge, MA: Harvard University Press.

Leach, Edmund. 1964. Anthropological Aspects of Language: Animal Categories and Verbal Abuse. In E. H. Lenneberg, ed. *New Directions in the Study of Language*. Cambridge, MA: MIT Press. pp. 23–63.

Lears, T. J. Jackson. 1981. *No Place of Grace: Antimodernism and the Transformation of American Culture 1880–1920*. Chicago: University of Chicago Press.

———. 1994. *Fables of Abundance: A Cultural History of Advertising in America*. New York: Basic Books.

Lepawsky, Joshua. 2014. Are We Living in a Post-Basel World? *Area*. http://onlinelibrary.wiley.com/journal/10.1111/(ISSN)1475-4762/earlyview.

Levi-Strauss, Claude. 1955. The Structural Study of Myth. *Journal of American Folklore* 68(270): 428–44.

Lewis, Kenneth. 2002. *West to Far Michigan: Settling the Lower Peninsula, 1815–1860*. East Lansing: Michigan State University Press.

Little, Peter. 2014. *Toxic Town: IBM, Pollution and Industrial Risks*. New York: NYU Press.

Logan, William B. 1995. *Dirt: The Ecstatic Skin of the Earth*. New York: W.W. Norton.

Lopata, Helena Z. 1993. The Interweave of Public and Private: Women's Challenge to American Society. *Journal of Marriage and Family* 55(1): 176–90.

MacBride, Samantha. 2012. *Recycling Reconsidered: The Present Failure and Future Promise of Environmental Action in the United States*. Cambridge, MA: MIT Press.

MacKenzie, Donald. 2009. *Material Markets: How Economic Actors Are Constructed*. Oxford, UK: Oxford University Press.

MacLeish, Kenneth T. 2012. Armor and Anesthesia: Exposure, Feeling, and the Soldier's Body. *Medical Anthropology Quarterly* 26(1): 49–68.

Macpherson, C. B. 1962. *The Political Theory of Possessive Individualism: Hobbes to Locke*. Oxford, UK: Clarendon Press.

Marres, Noortje. 2012. *Material Participation: Technology, the Environment, and Everyday Publics*. New York: Palgrave Macmillan.

Marriott, McKim and Ronald B. Inden. 1977. Toward an Ethnosociology of South Asian Caste Systems. In K. David, ed. *The New Wind: Changing Identities in South Asia*. Paris: Mouton. pp. 227–38.

Martin, Emily. 1987. *The Woman in the Body: A Cultural Analysis of Reproduction*. Boston: Beacon Press.

Mascarenhas, Michael. 2007. Where the Waters Divide: First Nations, Tainted Water and Environmental Justice in Canada. *Local Environment* 12(6): 565–77.

Masco, Joseph. 2006. *Nuclear Boderlands: The Manhattan Project in Post-Cold War New Mexico*. Princeton, NJ: Princeton University Press.

Maurer, Bill. 2005. *Mutual Life, Limited: Islamic Banking, Alternative Currencies, Lateral Reason*. Princeton, NJ: Princeton University Press.

McClurken, James M. 1986. Ottawa Adaptive Strategies to Indian Removal. *Michigan Historical Review* 12(1): 29–55.

McDougall, Allan K. and Lisa Philips. 2012. The State, Hegemony and the Historical British–US Border. In T. M. Wilson and H. Donnan, eds. *A Companion to Border Studies*. London: Wiley-Blackwell. pp. 177–93.

McGrady, David. 2006. *Living with Strangers: The Nineteenth-Century Sioux and the Canadian–American Borderlands*. Lincoln: University of Nebraska Press.

Medina, Martin. 2000. Scavenger Cooperatives in Asia and Latin America. *Resources, Conservation and Recycling* 31(1): 51–69.

Mellström, Ulf. 2004. Machines and Masculine Subjectivities: Technology as an Integrated Part of Men's Life Experiences. *Men and Masculinities* 6(4): 368–82.

Millar, Kathleen. 2008. Making Trash into Treasure: Struggles for Autonomy on a Brazilian Garbage Dump. *Anthropology of Work Review* 29(2): 25–34.

———. 2012. Trash Ties: Urban Politics, Economic Crisis and Rio de Janeiro's Garbage Dump. In C. Alexander and J. Reno, eds. *Economies of Recycling: Transformations of Materials, Values and Social Relations*. London: Zed Books. pp. 164–85.

Miller, Bruce G. 2006. Conceptual and Practical Boundaries: West Coast Indians/First Nations on the Border of Contagion in the Post-9/11 Era. In S. Evans, ed. *The Borderlands of the American and Canadian Wests: Essays on Regional History of the Forty-ninth Parallel*. Lincoln: University of Nebraska Press. pp. 49–66.

Miller, Daniel. 1998. *A Theory of Shopping*. Cambridge, UK: Polity Press.

———. 2012. *Consumption and Its Consequences*. Cambridge, UK: Polity Press.

Miller, Daniel and Heather Horst. 2013. The Digital and the Human: A Prospectus for Digital Anthropology. In D. Miller and H. Horst, eds. *Digital Anthropology*. London: Berg. pp. 3–38.

Miller, Ian W. 1997. *The Anatomy of Disgust*. Cambridge, MA: Harvard University Press.

Mitchell, Stanley. 1974. From Shklovksy to Brecht: Some Preliminary Remarks towards a History of the Politicization of Russian Formalism. *Screen* 15(2): 74–81.

Mol, Annemarie. 2002. *The Body Multiple: Ontology in Medical Practice*. Durham, NC: Duke University Press.

Molleson, Theya. 1981. The Archaeology and Anthropology of Death: What the Bones Tell Us. In S. C. Humphreys and H. King, eds. *Mortality and Immortality: The Anthropology and Archaeology of Death*. London: Academic Press. pp. 15–32.

Moore, Henrietta. 1986. *Space, Text, and Gender: An Anthropological Study of the Marakwet of Kenya*. New York: Guilford Press.

Mueggler, Erik. 2005. The Lapponicum Sea: Matter, Sense, and Affect in the Botanical Exploration of Southwest China and Tibet. *Comparative Studies in Society and History* 47(3): 442–79.

Muniesa, Fabian, Yuval Millo, and Michel Callon. 2007. *Market Devices*. London: Blackwell Wiley.

Nadasdy, Paul. 2003. *Hunters and Bureaucrats: Power, Knowledge, and Aboriginal–State Relations in the Southwest Yukon*. Vancouver: University of British Columbia Press.

Nagle, Robin. 2013. *Picking Up*. New York: Farrar, Straus and Giroux.

Neveu, Lily P. 2010. Beyond Recognition and Coexistence: Living Together. In L. Davis, ed. *Alliances: Re/Envisioning Indigenous–Non-Indigenous Relationships*. Toronto: University of Toronto Press. pp. 234–55.

Niezen, Ronald. 2003. *The Origins of Indigenism: Human Rights and the Politics of Identity*. Berkeley: University of California Press.

Nye, David E. 2003. *America as Second Creation: Technology and Narratives of New Beginnings*. Cambridge, MA: MIT Press.

O'Brien, Martin. 2007. *A Crisis of Waste? Understanding the Rubbish Society*. London: Routledge.

O'Neill, Kate. 2000. *Waste Trading among Rich Nations: Building a New Theory of Environmental Regulation*. Cambridge, MA: MIT Press.

Ong, Aihwa. 2006. *Neoliberalism as Exception: Mutations in Citizenship and Sovereignty*. Durham, NC: Duke University Press.

Osborne, Thomas. 1996. Security and Vitality: Drains, Liberalism and Power in the Nineteenth Century. In A. Barry, T. Osborne, and N. Rose, eds. *Foucault and Political Reason: Liberalism, Neo-liberalism and Rationalities of Government*. Chicago: University of Chicago Press. pp. 99–122.

Packard, Vance. 1960. *The Waste Makers*. New York: David McKay.

Palley, Thomas. 2011. America's Flawed Paradigm: Macroeconomic Causes of the Financial Crisis and Great Recession. *Empirica* 38: 3–17.

Parmentier, Richard. 1979. The Mythological Triangle: Poseyemu Montezuma and Jesus in the Pueblos. In W. Sturtevant, ed. *Handbook of North American Indians, vol. 9: Southwest*. Washington, DC: Smithsonian Institution.

Parry, Jonathan. 1982. Sacrificial Death and the Necrophagous Ascetic. In J. Parry and M. Bloch, eds. *Death and the Regeneration of Life*. New York: Cambridge University Press. pp. 74–110.

———. 1994. *Death in Banaras*. Cambridge, UK: Cambridge University Press.

Parsons, Elsie C. 1980 [1925]. *The Pueblo of Jemez*. New York: AMS Press.

Patton, Paul and John Protevi, eds. 2003. *Between Deleuze and Derrida*. London: Bloomsbury Academic.

Pedersen, David. 2008. Brief Event: The Value of Getting to Value in the Era of 'Globalization.' *Anthropological Quarterly* 8(1): 57–77.

———. 2013. *American Value: Migrants, Money, and Meaning in El Salvador and the United States.* Chicago: University of Chicago Press.

Pellow, David N. 2007. *Resisting Global Toxics.* Cambridge, MA: MIT Press.

Perry, Stewart E. 1998. *Collecting Garbage: Dirty Work, Clean Jobs, Proud People.* New Brunswick, NJ: Transaction.

Pietz, William. 1985. The Problem of the Fetish, I. *RES: Anthropology and Aesthetics* 9: 5–17.

———. 1987. The Problem of the Fetish, II. *RES: Anthropology and Aesthetics* 13: 23–45.

Piot, Charles. 2010. *Nostalgia for the Future: West Africa after the Cold War.* Chicago: University of Chicago Press.

Porter, Dorothy. 1999. *Health, Civilization and the State: A History of Public Health from Ancient to Modern Times.* New York: Routledge.

Postone, Moishe. 1993. *Time, Labor, and Social Domination: A Reinterpretation of Marx's Critical Theory.* Cambridge, UK: Cambridge University Press.

Povinelli, Elizabeth. 2006. *The Empire of Love: Toward a Theory of Intimacy, Genealogy, and Carnality.* Durham, NC: Duke University Press.

Protevi, John. 2009. *Political Affect.* Minneapolis: University of Minnesota Press.

———. 2011. Ontology, Biology, and History of Affect. In L. Bryant, N. Srnicek, and G. Harman, eds. *The Speculative Turn: Continental Materialism and Realism.* Melbourne: re.press. pp. 393–405.

———. 2013. *Life, War, Earth.* Minneapolis: University of Minnesota Press.

Rabinow, Paul. 1996. *Essays on the Anthropology of Reason.* Princeton, NJ: Princeton University Press.

Rahtz, Philip. 1981. Artefacts of Christian Death. In S. C. Humphreys and H. King, eds. *Mortality and Immortality: The Anthropology and Archaeology of Death.* London: Academic Press. pp. 117–36.

Ramirez, Bruno. 2001. *Crossing the 49th Parallel: Migration from Canada to the United States, 1900–1930.* Ithaca, NY: Cornell University Press.

Rappaport, Roy. 1999. *Ritual and Religion in the Making of Humanity.* Cambridge, UK: Cambridge University Press.

Rasmussen, Joel. 2010. Enabling Selves to Conduct Themselves Safely: Safety Committee Discourse as Governmentality in Practice. *Human Relations* 64(3): 459–78.

———. 2013. Governing the Workplace or the Worker? Evolving Dilemmas in Chemical Professionals' Discourse on Occupational Health and Safety. *Discourse & Communication* 7(1): 75–94.

Rathje, William and Cullen Murphy. 2001. *Rubbish! The Archaeology of Garbage.* New York: HarperCollins.

Reid, Donald. 1991. *Paris Sewers and Sewermen: Realities and Representations.* Cambridge, MA: Harvard University Press.

Reid, J. Jefferson, Michael B. Schiffer, and William L. Rathje. 1975. Behavioral Archaeology: Four Strategies. *American Anthropologist* 77(4): 864–9.

Reno, Joshua. 2009. Your Trash Is Someone's Treasure: The Politics of Value at a Michigan Landfill. *Journal of Material Culture* 14(1): 29–46.

———. 2012. Evidence Excess: Material Deposits and Narcotics Surveillance in the USA. In C. Alexander and J. Reno, eds. *Economies of Recycling: The Global Transformation of Materials, Values and Social Relations.* London: Zed Books. pp. 234–54.

———. 2014. Toward a New Theory of Waste: From 'Matter out of Place' to Signs of Life. *Theory, Culture & Society* 31(6): 3–27.

———. 2015. Waste and Waste Management. *Annual Review of Anthropology* 44: in press.

Roediger, David. 1999. *The Wages of Whiteness: Race and the Making of the American Working Class.* New York: Verso.

Rogers, Heather. 2005. *Gone Tomorrow: The Hidden Life of Garbage.* New York: New Press.

Rose, Nikolas. 1999. *Powers of Freedom: Reframing Political Thought.* Cambridge, UK: Cambridge University Press.

———. 2007. *The Politics of Life Itself: Biomedicine, Power, and Subjectivity in the Twenty-first Century.* Princeton, NJ: Princeton University Press.

Rosen, Christine M. 1997. Noisome, Noxious, and Offensive Vapors, Fumes and Stenches in American Towns and Cities, 1849–1865. *Historical Geography* 25: 49–82.

Rotman, Brian. 1987. *Signifying Nothing: The Semiotics of Zero.* London: Macmillan.

Royte, Elizabeth. 2006. *Garbage Land: On the Secret Trail of Trash.* Boston: Beacon Press.

Rozin, Paul, Linda Millman, and Carol Nemeroff. 1986. Operation of the Laws of Sympathetic Magic in Disgust and Other Domains. *Journal of Personality and Social Psychology* 50(4): 703–12.

Rozyman, Edward B. and John Sabini. 2001. Something It Takes to Be an Emotion: The Interesting Case of Disgust. *Journal for the Theory of Social Behaviour* 31(1): 29–59.

Sahlins, Marshall. 1968. Notes on the Original Affluent Society. In R. B. Le and I. De Vore, eds. *Man the Hunter.* Chicago, IL: Aldine. pp. 85–9.

———. 2004. *Apologies to Thucydides: Understanding History as Culture and Vice Versa.* Chicago: University of Chicago Press.

Salamon, Sonya. 2007. *Newcomers to Old Towns: Suburbanization of the Heartland.* Chicago: University of Chicago Press.

Sanger, Matthew. 2015. Life in the Round: Social and Ritual Lives of the Southeastern Coastal Late Archaic. PhD dissertation, Department of Anthropology, Columbia University, New York.

Schiffer, Michael. 1972. Archaeological Context and Systemic Context. *American Antiquity* 37: 156–65.

———. 1987. *Formation Processes of the Archaeological Record.* Albuquerque: University of New Mexico Press.

Schrödinger, Erwin. 1944. *What Is Life?* Cambridge, UK: Cambridge University Press.

Sennett, Richard and Jonathan Cobb. 1993. *Hidden Injuries of Class.* New York: W.W. Norton.

Serres, Michel. 1982. *The Parasite.* Baltimore: Johns Hopkins University Press.

———. 2010. *Malfeasance: Appropriation through Pollution?* Palo Alto, CA: Stanford University Press.

Shanks, Michael, David Platt, and William Rathje. 2004. The Perfume of Garbage: Modernity and the Archaeological. *Modernism/modernity* 11(1): 61–83.

Shapiro, Nicholas. 2015. Attuning to the Chemosphere: Domestic Formaldehyde, Bodily Reasoning, and the Chemical Sublime. *Cultural Anthropology* 30(3): 368–393.

Silko, L. M. 1986. Landscape, History, and the Pueblo Imagination. In D. Halpern, ed. *On Nature—Nature, Landscape, and Natural History.* San Fransisco: North Point Press. pp. 83–94.

Simmons, Phil, Nora Goldstein, Scott M. Kaufman, Nickolas J. Themelis, and James Thompson, Jr. 2006. The State of Garbage in America. *Biocycle* 47(4): 26–43.

Sjöberg, Lennart. 2004. Local Acceptance of a High-Level Nuclear Waste Repository. *Risk Analysis* 24(3): 737–49.

Slotkin, Richard. 1985. *The Fatal Environment: The Myth of the Frontier in the Age of Industrialization 1800–1890.* New York: Antheneum.

Slowey, Gabrielle. 2008. *Navigating Neoliberalism: Self-Determination and the Mikisew Cree First Nation.* Vancouver: Farrar, Straus and Giroux.

Smith, Barbara H. 1988. *Contingencies of Value: Alternative Perspectives for Critical Theory.* Cambridge, MA: Harvard University Press.

Smith, Henry N. 1950. *Virgin Land: the American West as Symbol and Myth.* Cambridge, MA: Harvard University Press.

Smith, Neil. 1990. *Uneven Development: Nature, Capital, and the Production of Space.* Oxford, UK: Blackwell.

———. 2003. *American Empire: Roosevelt's Geographer and the Prelude to Globalization.* Berkeley: University of California Press.

Soldatic, Karen. 2011. Appointment Time: Disability and Neoliberal Workfare Temporalities. *Critical Sociology* 39(3): 405–19.

Sørensen, Tim F. 2009. The Presence of the Dead: Cemeteries, Cremation and the Staging of Non-Place. *Journal of Social Archaeology* 9(1): 110–35.

Spyer, Patricia. 1998. Introduction. In P. Spyer, ed. *Border Fetishisms: Material Objects in Unstable Spaces.* New York: Routledge. pp. 1–12.

Stallybrass, Peter. 1999. Marx's Coat. In P. Spyer, ed. *Border Fetishisms: Material Objects in Unstable Spaces.* New York: Routledge. pp. 183–207.

Stallybrass, Peter and Allon White. 1986. *The Politics and Poetics of Transgression.* New York: Cornell University Press.

Staten, Henry. 1984. *Wittgenstein and Derrida.* Lincoln: University of Nebraska Press.

Stewart, Kathleen. 1996. *A Space on the Side of the Road: Cultural Poetics in an "Other America.* Princeton, NJ: Princeton University Press.

Stoler, Ann, ed. 2006. *Haunted by Empire: Geographies of Intimacy in North American History.* Durham, NC: Duke University Press.

———, ed. 2007. *Imperial Formations.* Santa Fe, NM: School for Advanced Research Press.

———, ed. 2013. *Imperial Debris: On Ruins and Ruination.* Durham, NC: Duke University Press.

Strasser, Susan. 1999. *Waste and Want: A Social History of Trash.* London: Metropolitan Books.

Strathern, Andrew. 1981. Death as Exchange: Two Melanesian Cases. In S. C. Humphreys and H. King, eds. *Mortality and Immortality: The Anthropology and Archaeology of Death.* London: Academic Press. pp. 205–24.

Strathern, Marilyn. 1988. *The Gender of the Gift.* Berkeley: University of California Press.

———. 2004. *Partial Connections.* Walnut Creek, CA: AltaMira Press.

Sugrue, Thomas. 1996. *The Origins of the Urban Crisis.* Princeton, NJ: Princeton University Press.

Swaney, James A. 1994. So What's Wrong with Dumping on Africa? *Journal of Economic Issues* 28: 367–77.

Taussig, Michael. 1989. History as Commodity in Some Recent American (Anthropological) Literature. *Critique of Anthropology* 9(1): 7–23.

Teaford, Jon C. 1993. *Cities of the Heartland: The Rise and Fall of the Industrial Midwest.* Bloomington: Indiana University Press.

Thompson, E. P. 1967. Time, Work-Discipline and Industrial Capitalism. *Past and Present* 38(1): 56–97.

Thomson, Vivian. 2009. *Garbage In, Garbage Out: Solving the Problems with Long-Distance Waste Transport.* Charlottesville: University of Virginia Press.

Thorpe, Grace. 1996. Our Homes Are Not Dumps: Creating Nuclear Free Zones. *Natural Resources Journal* 36: 955–63.

Tomes, Nancy. 1999. *The Gospel of Germs: Men, Women, and the Microbe in American Life.* Cambridge, MA: Harvard University Press.

Trouillot, Michel-Rolph. 1995. *Silencing the Past: Power and the Production of History.* New York: Beacon Press.

Tsing, Anna. 2005. *Friction: An Ethnography of Global Connection.* Princeton, NJ: Princeton University Press.

Twigg, Julia, Carol Wolkowitz, Rachel L. Cohen, and Sarah Nettleton. 2011. Conceptualizing Body Work in Health and Social Care. *Sociology of Health & Wellness* 33(2): 171–88.

Urciuoli, Bonnie. 1996. *Exposing Prejudice: Puerto Rican Experiences of Language, Race, and Class.* Boulder, CO: Westview Press.

Veblen, Thorstein. 1899. *The Theory of the Leisure Class.* New York: Macmillan.

Vizenor, Gerald. 1991. *Landfill Meditation: Crossblood Stories.* Hanover, NH: University Press of New England.

Walker, William H. and Lisa J. Lucero. 2000. The Depositional History of Ritual and Power. In M. A. Dobres and J. Robb, eds. *Agency in Archaeology*. London: Routledge.

Wallace, David F. 1996. *Infinite Jest*. Boston: Back Bay Books/Little, Brown.

Walsh, Edward. 1975. *Dirty Work, Race, and Self-Esteem*. Ann Arbor: University of Michigan, Institute for Industrial and Labor Relations.

Wander, Katherine, Kathleen O'Connor, and Bettina Shell-Duncan. 2012. Expanding the Hygiene Hypothesis: Early Exposure to Infectious Agents Predicts Delayed-Type Hypersensitivity to *Candida* among Children in Kilimanjaro. *PLoS ONE* 7(5): e37406.

Warren, Kim. 2007. Separate Spheres: Analytical Persistence in United States Women's History. *History Compass* 5(1): 262–77.

Watts, Edward. 2001. *An American Colony: Regionalism and the Roots of Midwestern Culture*. Athens: Ohio University Press.

Weiner, Annette. 1992. *Inalienable Possessions: The Paradox of Keeping-While-Giving*. Berkeley: University of California Press.

White, Richard. 1991. *The Middle Ground: Indians, Empires, and Republics in the Great Lakes Region, 1650–1815*. Cambridge, UK: Cambridge University Press.

Williams, William A. 1958. The Age of Mercantilism. *William and Mary Quarterly* 15: 419–37.

Yates, Michelle. 2011. The Human-As-Waste, the Labor Theory of Value and Disposability in Contemporary Capitalism. *Antipode* 43(5): 1679–95.

Zimring, Carl. 2004. Dirty Work: How Hygiene and Xenophobia Marginalized the American Waste Trades, 1870–1930. *Environmental History* 9(1): 80–101.

———. 2009. *Cash for Your Trash: Scrap Recycling in America*. Brunswick, NJ: Rutgers University Press.

Žižek, Slavoj. 2010. *Living in the End Times*. London: Verso.

incineration, 13, 63, 96, 213–14; compared
with landfills, 28–29, 133, 223. *See also*
infrastructure; sanitary engineering;
waste

indigenous people. *See* Native Americans;
Pueblo; Timiskaming

individualism, 40, 61, 68, 72, 84–86, 92,
124, 134, 172–73, 178, 214, 217–18, 222;
self-creation and, 14, 15–16, 25–26, 60,
76–79, 132–3; settler colonialism and,
137–8, 196–97. *See also* character; family; personhood; transcendence

infrastructure, 138–39, 175; transportation,
22–23, 34–35, 44–45, 47, 53, 136, 138, 140,
146–52, 157–58, 161, 178–81, 183–87,
204–7, 209–10, 226; waste management,
7, 12–13, 18, 25, 132, 212–17, 222–24. *See
also* energy; landfill; sanitary engineering

Ingold, Tim, 20, 242n4

labor. *See* value; waste work

landfill, 1–5; as social relationship, 2, 12, 14,
16, 23, 28, 54, 57, 60, 99, 103, 133, 136–37,
165, 173, 207–8, 214–15, 218, 222; burial
of people within, 218–20, 223–24;
disposability and, 6, 8, 10–11, 16–17, 57,
101, 103–9, 133–34, 215, 227; leakiness of,
22–57, 61, 152, 225; modernity and, 6,
12–16, 22–23, 29, 31–32, 54, 56, 103, 207,
216; reclamation of, 126, 219, 225–27. *See
also* body; sanitary engineering; waste;
waste market; waste work

landscape, 6–7, 115–16, 137, 139, 145–46,
150–51, 153, 155–56, 162–63, 165–67, 220,
224; immersion within, 26, 32–39,
145–46. *See also* emplacement;
wasteland

language. *See* idioms; semiotics; speech acts

liberalism, 25, 39, 131, 203. *See also* body;
Locke, John; neoliberalism

lines: ghostly, 171–72, 242n4; made concrete, 173–74, 188–93, 193–95. *See also*
borders; race

Locke, John, 131–33, 241n13

Love Canal disaster, 23

market, 75, 93–94, 95, 103–9, 115, 120, 121,
125, 132–33, 147, 174; internal labor

market, 64, 85. *See also* economics;
two-sphere ideology; waste market

Marx, Karl, 65–6, 111, 133–34. *See also*
capital; fetishism; value

masculinity, 111–12; fatherhood, 9, 39,
69–84, 111–12, 115, 128, 159–60, 202;
labor and, 15, 58, 86–95; scavenging and,
102, 113–17, 120–21, 134–35; sociability
and, 43–44, 112–15, 126–31. *See also*
gender; family; tinkering; two-sphere
ideology

mass consumption, 2, 96, 98–103, 104–5,
108–12, 118, 131–32; as a choice, 99–100,
103, 133–34, 212. *See also* commodities;
market; mass waste; shopping; two-
sphere ideology

materiality. *See* absence; cybernetics;
entropy; form; ontology; semiotics

memory, 46, 54, 75, 143, 153–57, 159–61, 186;
lost, 10, 145–48, 150–51, 154; objects of,
94, 114–15, 118–19, 122–23, 137, 165–66,
217, 219–20, 223. *See also* emplacement;
historicity

methane, 28–29, 33, 38, 51–52, 85, 90–91,
125. *See also* energy; microbes; odor

miasmas, 54–55. *See also* odor; sanitary
engineering

Michigan, 1, 15, 132. 159, 183–84; auto
industry, 71–72, 80, 110–11, 121, 177–78;
football, 83; geology, 158; history, 146–
50, 154–55, 157, 160, 165–66, 178, 195;
regulators, 31–32, 164, 188–93; waste
importation by, 168–174, 177–82, 207,
213. *See also* North America; United
States; waste market

microbes, 26–27, 38, 51–52, 54, 55; germs,
23, 69. *See also* energy; methane

middle-class, 2, 98, 126, 142, 213; aspirations, 15, 16, 39, 48, 50, 59–61, 67–68,
71–72, 77–78, 82–84, 94–97, 111, 115,
129, 131, 135, 160. *See also* class; family;
houses; transcendence

money, 58–61, 92, 100, 165; landfill-host
agreements, 162–63, 225; price, 51,
62–64, 104, 112, 174–75; scavenged,
121–24, 134; spoilage and, 95–97, 131–33;
wages, 15, 39, 48, 59–60, 64–67, 70–82,
84–87, 92–94, 99, 102, 110–12, 115–17,